D1422982

The Economics of Pig Production

*Records and performance of pig production
as monitored by
The Cambridge Pig Management Scheme*

BOB RIDGEON

Moulton College

Cat. No. 636.4

Acn. No. 11484

DB

FARMING PRESS

First published 1993

Copyright © Bob Ridgeon 1993

All rights reserved. No parts of this
publication may be reproduced, stored in
a retrieval system, or transmitted, in any
form or by means, electronic, mechanical,
photocopying, recording or otherwise, without
prior permission of Farming Press Limited.

ISBN 0 85236 269 2

A catalogue record for this book is available
from the British Library

Published by Farming Press Books
Wharfedale Road, Ipswich IP1 4LG, United Kingdom

Distributed in North America
by Diamond Farm Enterprises,
Box 537, Alexandria Bay, NY 13607, USA

Cover design by Andrew Thistlewaite
Front cover photograph courtesy of Pig Farming
Typeset by Carol Asby and Joy Meyrick
Printed and bound in Great Britain by
Butler & Tanner,
Frome and London

Contents

Acknowledgements

University departments of agricultural economics have for many years undertaken economic studies of farming activities, receiving financial and technical support from the Ministry of Agriculture, Fisheries and Food. Acknowledgement is gratefully expressed for the benefit derived by the Cambridge Pig Management Scheme.

No survey requiring detailed information would be possible without the willing participation of the co-operators. Special thanks are due to those farmers who regularly completed records of their pig units despite growing pressures on their time. They have contributed to our knowledge and allowed us to monitor the often rapidly changing pig business to provide the industry and policy makers with important facts and figures. Over the years I have received much information and advice from many people and organisations, including individual members of the feed and meat trades, too numerous to mention, and this help has been greatly appreciated.

I have a deep sense of gratitude to Dr A W Menzies-Kitchin, the director of the Farm Economics Branch of the School of Agriculture until 1951, for his pioneering work in pig recording and for providing me with employment in this University. I am indebted also to his successor, Ford Sturrock, for encouraging my work with pigs and for entrusting me with the responsibility of running a field survey. Thanks are particularly due to Ian Sturgess, the current director (now the Agricultural Economics Unit), for his continuous support and the suggestion that I write this book. He also found time during a busy sabbatical to read the manuscript and comment on the content and readability. I am extremely grateful to John Hill, a close associate for many years, for his stimulating encouragement in this work and for reading the manuscript and making valuable suggestions.

I am also indebted to many colleagues for invaluable help and guidance given to me over the years and in the production of this book. Eleanor Lyons for 20 years competently assisted in the scrutiny and computation of records and her successor, David Wilkins, developed and continually upgraded systems for processing data by personal computer. Since 1971 the scheme results have been calculated by mainframe computer and thanks are due to Faisal Sabbah for maintaining the programs over many years. I would like to express my sincere appreciation to Geoff Davidson and John Squirrell for coming to the aid of this ageing field worker in recent years by undertaking some of the farm visits. Thanks are due to Joy Meyrick for the laborious work of typing drafts and the preparation of camera-ready copy for the printers. I owe a considerable debt of gratitude to Carol Asby for the graphical presentations and for her production skills in liaison with Farming Press Books in the preparation of this book for publication.

Bob Ridgeon   September 1993
Cambridge

Foreword

During the past forty years the Cambridge Pig Management Scheme has been the main fount of economic data at the farm level on pig production in the United Kingdom. Its reports have informed policy debates and provided the numerical sinews of the advice of management consultants, bankers and accountants. There have been other sources and increasingly so but few of these have been seen to be as objective, meticulous, comprehensive, apposite or timely.

The high reputation of the scheme owes much to the single-minded, and in many respects single-handed, input of Bob Ridgeon. Throughout this period he has done almost all the fieldwork, scrutinised data with an eagle eye and master-minded the analysis. His annual reports have been mainly numerical. The tables were carefully designed to speak for themselves and modified in response to changing technology, policy and markets. However, long and close contacts with pig producers, together with a shrewd understanding of management, have given Bob a bigger story to tell. Episodes and aspects of this story have been presented to select audiences and readerships but have never been put together in a comprehensive and accessible way. Bob Ridgeon has devoted his last few months before retirement to putting this right.

The end result is a book which is at once readable and rigorous. The determinants of the profitability of pig production are presented precisely and in their historical context. It provides a valuable source of perspective and reference for those who are concerned with pig production, directly or indirectly, or who aspire to be so.

Ian Sturgess   September 1993
Director
Agricultural Economics Unit
Department of Land Economy
University of Cambridge

Chapter 1

Introduction

Standards of pig production have improved most impressively during the second half of the twentieth century. This progress has been monitored in detail by the Cambridge Pig Management Scheme. It shows that had pigs been produced in the early 1990s at the same level of efficiency in feed requirements as in 1950, then no pig keepers would have made any money. Costs, excluding interest charges, would have averaged 142p per kg deadweight, far in excess of the 109p per kg received for pigs sold in the three years 1990-92.

Advances in technology have enabled performance to improve substantially. On average, the number of pigs produced increased from 12.8 in 1950 to 21 per sow in the 1990s and the quantity of feed used has been reduced from 6.6 kg to 4.1 kg per kg deadweight of pigmeat.

The rewards from pig-keeping have not been over-generous, especially since 1974. Most of the savings from increased efficiency would appear to have been passed on to consumers by containing retail prices. In fact, to maintain the level of income from pigs, most producers have found it necessary to aim continually for improvements and often expand their herds. Failure to keep pace with progress seems a common reason why many have given up pigs.

Most successful pig producers manage their herds as a business and record with care. They are kind to their animals and, by using appropriate housing and suitable feed, they provide a good environment to ensure a profitable unit. Results from their records should indicate any weaknesses where there is room for improvement, so that attention can be directed to those areas to rectify matters.

History of the Cambridge Pig Management Scheme

With over 20 per cent of the United Kingdom's pigs in the Eastern region, it is perhaps not surprising that the study of pig production has featured strongly in agricultural economics at Cambridge for many years. Pig recording via the East Anglian Pig Recording Scheme was first undertaken at Cambridge in 1928 by the then University Department of Agriculture and dealt mainly with assessing sow and litter performance. The aim was to correct the complete absence of the necessary records of achievements in Great Britain by encouraging pig producers to overcome their apparent inability, or unwillingness, to keep figures which could be used to measure and compare the standards of individual sows in the herd. From these results a start was made on compiling suitable standards of production to form a basis for comparing the efficiency of British producers. Such records also provided indicators of how the farmer's management of his pig herd was succeeding or failing.

Later the emphasis moved from sow and litter performance, where perhaps only the best sows were included, to recording the whole herd. This allowed more satisfactory measurements of overall achievement. In 1936, a survey to analyse the economic factors affecting the profitability of pig production and to lay down standards of herd performance for pig producers was established by the Farm Economics Branch (later to become the Agricultural Economics Unit). This investigation was the forerunner of the Cambridge Pig Management Scheme which, apart from the 1939-45 war years, operated continuously until 1991. The scheme developed over the years to reflect the intricacies of technical advances and changes brought about by increasing herd sizes and specialisation. Initially the scheme was a basic costs and returns study on approximately 50 small herds by current standards, it moved on to a detailed investigation of physical and financial aspects of pig production on about 150 much larger and randomly selected herds. The scheme thus increased considerably in size and complexity. For many years it has been widely recognised as the main source of reliable economic data on pig production, with a high reputation in Government and academic circles at home and abroad and all sectors of the pig industry.

Throughout its existence the Cambridge Pig Management Scheme received funding via the Economics Division of the Ministry of Agriculture, Fisheries and Food, but was allowed independence and freedom of action and comment. In 1992, however, the Ministry decided to replace the continuous Cambridge study, together with the similar one at Exeter, with the single National Pig Survey to run for two in every five years. The stated aim of this change was to make best use of the resources available, so that the coverage of other livestock and crops could be improved and at the same time to extend regional coverage of the pig sample beyond the Eastern Counties. Other pig producing areas had not been covered previously except by small surveys of brief duration at other university recording centres in the 1960s.

The Ministry has a greater involvement in the planning of the National Pig Survey and as its requirements are now more limited, some simplification has been introduced to ease the task of all participants. The modified survey, which is being co-ordinated at the University of Exeter, runs for 1992 and 1993 on a trial basis, with the combined records mainly from the existing Cambridge and Exeter herd samples, supplemented by small numbers from the South of England and Yorkshire. When the National Pig Survey starts the full investigation in 1996 it is planned that all regions will participate, each with a sample of herds to reflect the number of pigs in each region.

Objectives

The objectives of the Cambridge Pig Management Scheme were to monitor representative commercial herds to provide up to date information on the economics of pig production, including quantitative standards of performance and costs involved, together with year by year changes, for use in policy assessments by Government, farmers and others associated with the pig industry.

For many years replacement herds in the scheme have been randomly selected from lists of pig producers supplied in confidence by the Ministry from the agricultural census. The aim has been to cover all systems and types of production for all farm and herd sizes. Co-operators were required to complete monthly details of stock purchases and sales, births and deaths and the feed used. Valuations of stock on hand, together with labour and other costs, were collected on half-yearly visits. To ensure accuracy all records were thoroughly scrutinised for feasibility and anything suspect was queried. Only then were records consid-

ered valid enough to give a fair reflection of what had actually
occurred on the farm. Any unreliable records were omitted from
the results.

Co-operation with producers

Usually, it was not difficult to persuade the selected pig producers to take part in the scheme, though occasionally herds in certain categories were hard to find. Most had previously heard
about the scheme and felt its existence was beneficial to producers and the pig industry. The aim was to keep records as simple
as possible, with the monthly details on just one sheet of paper
and designed so that all pigs were accounted for before sending
in the return for processing.

Co-operation with pig producers evolved over several
years and, with such a long-standing survey, there was a willing-
ness to answer any additional questions about their business for
supplementary investigations. Recent developments and topical
issues always stimulated considerable interest and a keenness to
help. It is unlikely that such a good working relationship, built up
over several years from knowledge of the farm in question, could
have been established by intermittent short surveys.

The provision of prompt and accurate feedback is an
essential part of collecting worthwhile and reliable data on a
regular basis from voluntarily co-operating farmers. In return for
supplying this information, each farmer in the scheme received
the results of his own herd every six months in the form of a
trading account and a list of production performance factors
achieved. Each co-operator also received annual results from an
amalgamation of the two six-monthly periods.

Results kept simple

The results were presented in a straightforward manner which
could be readily understood. Many yardsticks are employed in
pig recording but the Cambridge experience showed that a greater impact could be made by using a select few of the more important factors. The simpler and more objective the analysis, the
more busy farmers and their staff are likely to derive benefit
from the information. Too many factors often contain similari-

4

ties or repeats and can confuse and distract attention from the key issues. If these are presented satisfactorily, the rest will usually fall into line. A comparison with average results from the scheme, where all herds are recorded in the same manner, will soon reveal the level of efficiency achieved and how it relates to others.

The importance of recording

While many pig producers take part in one of several schemes available, it is not uncommon to come across some who keep no records at all. The usual excuses are lack of time or insufficient interest. Producers often admit their dislike of book-keeping and confess to being disorganised. Some think that because they use good stockmen and know their pigs, there is no need to keep records. Others say the bank manager will soon contact them if the financial side of the business is failing, but by then, of course, it may be too late. With a pig unit on an arable or mixed farm, it can often be unclear whether one enterprise is supporting another. It is important to ascertain at least that the pig unit is not the passenger.

Size of herd is really irrelevant in justifying the need to record. Even a 50 sow herd producing bacon weight pigs could be expected to have an £80,000 turnover (1992 prices) and a profit margin of some £10,000, before interest charges, in a normal year. Without records and attention to detail, it is impossible to tell whether the herd performs satisfactorily or not. Most producers concede that the time spent recording the herd is amply rewarded by the information they receive in return about their own business and averages of others for comparison, which provides a basis for making sound management decisions.

Chapter 2

Structure of Pig Production

Pig numbers

Following the run-down of pig production during the war of 1939-45, numbers in the UK increased rapidly with Government encouragement until 1954. Then after a few years of relative stability increased further until the mid-1960s. Since then the total pig population has changed little, apart from a five year period in the early 1970s when a shortage of beef stimulated the demand for pigmeat, causing prices and numbers to rise. The record number of just under nine million was attained in 1973. When beef supplies recovered a year later, the profitability of pigs collapsed from over-production and numbers quickly declined to the previous level.

The ratio of total pigs (including the boars and sows) to breeding sows has increased over the years as sow productivity improved; see Figure 2.1. Until the mid-1960s, census results usually showed that there were about eight total pigs to one sow, though major expansions or contractions in numbers could briefly distort this relationship. The ratio gradually increased as earlier weaning became more popular and output per sow rose, reaching almost ten to one by the late 1980s. Growth rates have also improved but as slaughter weights have risen marginally in recent years, one possibly offsets the other, so that the time growing pigs remain on the farm is unlikely to have changed sufficiently to have much effect on the ratio.

The number of breeding sows in the UK peaked at just over one million in 1973. Around this time production amounted to 15 to 16 million slaughterings a year (excluding boars and sows), plus boars and gilts for breeding herd replacements and for export. More recently, slaughterings have been about 14

6

million pigs a year. Just under 800,000 sows have been enough to produce this level of output and rear replacement boars and gilts for breeding.

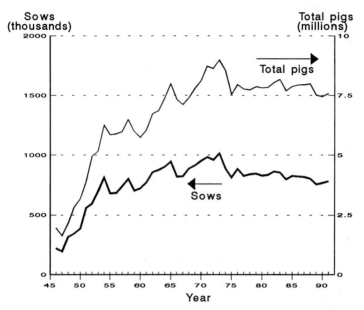

Figure 2.1 Breeding sow and total UK pig population

Location of pigs

Pig production has always been located predominantly in the cereal growing areas. The map in Figure 2.2 giving the distribution of pigs and cereals by university region[1] shows that this relationship still largely prevails. There are exceptions now, most notably in the North Eastern region of England, where the pig population has increased considerably, but this is partly compensated for by fewer pigs and more cereals in the neighbouring East

---

[1] University regions are used for agricultural surveys and differ from MAFF standard statistical regions by which census data are normally presented. The University Eastern region, for example, covered by the Cambridge Pig Management Scheme includes Norfolk, Suffolk, Essex, what was Holland division of Lincolnshire, Cambridgeshire, Bedfordshire, Hertfordshire and part of Greater London, whereas the MAFF region covers only Norfolk, Suffolk and Cambridgeshire.

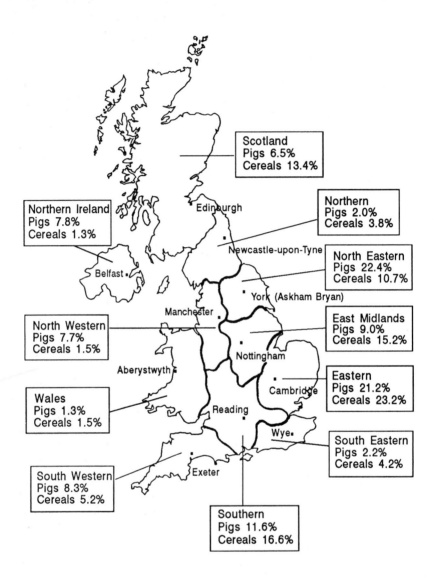

Source: Agricultural Census, MAFF

**Figure 2.2. Number of pigs and area of cereals 1991
(percentage distribution by region)**

Midlands region. The combined regions of the eastern side of England, on a line roughly from Yorkshire to the Isle of Wight, have between them 66 per cent of the UK pigs and grow 70 per cent of the cereals.

The North Eastern and the Eastern regions were the most densely populated pig areas in 1991, with 22 per cent and 21 per cent respectively of the UK pigs. Next came the Southern region with 11 per cent, followed by East Midlands (9 per cent) and South Western (8 per cent). The North Western region and Northern Ireland both had just under 8 per cent, while Scotland was down to 6.5 per cent. Wales and the Northern and South Eastern regions all had small populations of pigs.

Census results are often published for England and Wales separately from Scotland and Northern Ireland. Until the early 1970s, about 80 per cent of the pigs in the UK were in England and Wales. Since 1980 the share has been 86 per cent, mainly due to a decline in Scottish pig production.

Table 2.1 Breeding sows and total pigs by region 1991

Region	Sows			Total pigs		
	No. of hldgs	No. of sows	Av.No. per hldg	No. of hldgs	No. of pigs	Av.No. per hldg
Northern	295	16,639	56	432	149,224	345
North Eastern	1,770	177,497	100	2,376	1,697,328	714
East Midlands	773	69,427	90	1,103	685,342	621
Eastern	1,409	159,265	113	2,077	1,608,194	774
South Eastern	396	16,306	41	604	163,494	271
Southern	1,330	99,428	75	1,886	880,276	467
South Western	1,488	64,236	43	2,285	628,575	275
North Western	1,139	57,895	51	1,600	582,666	364
England	8,600	660,693	77	12,363	6,395,099	517
Wales	950	11,749	12	1,375	102,038	74
Eng & Wales	9,550	672,442	70	13,738	6,497,137	473
Scotland	546	51,375	94	902	493,025	547
N. Ireland	2,333	59,112	25	2,586	588,349	228
United Kingdom	12,429	782,929	63	17,226	7,578,511	440

Source: Agricultural Census, MAFF.

The distribution of breeding sows by region was similar to the distribution of total pigs. Northern Ireland had the highest number of holdings with sows (19 per cent of the UK total) and with total pigs (15 per cent), but with less than 8 per cent of sows and total pigs, the average number per holding was lower than all other regions except Wales. Table 2.1 shows that several other regions apart from the North Eastern and the Eastern regions had large numbers of holdings with pigs but the average numbers per herd were much smaller.

For sows, average herd sizes were greatest in the North Eastern and the Eastern regions and in Scotland. For total pigs, the Eastern region had on average the largest herds, followed by the North Eastern, the East Midlands and then Scotland. The smallest herds by far for both sows and for total pigs were Wales, then Northern Ireland, the South Eastern and the South Western regions.

The MAFF term "holdings" refers to completed census forms and caution should be exercised when using these figures. Many farms when acquiring land incorporate the addition, which may previously have had a separate census return, with their existing farm, or farms, but some do not and the addition remains a separate holding for the purpose of the census. Therefore, while most farmers complete only one census form, which may cover other farms in their name, a few will fill in two or more returns. A further possible complication when assessing the number of pig producers arises from the fact that any farmers keeping pigs for others, probably on a contract production scheme, have to include on their census form pigs they do not own. Whether these farmers can legitimately be classed as pig producers is subject to debate. As the farmers concerned have not been required to record their non-ownership on the census return, the number involved is unknown. Because of multiple returns and those not owning the pigs on their farm, the number of pig producers could reasonably be expected to be less than the number of holdings with pigs. Nevertheless, the census results still provide a good indication of the structure of pig production in this country.

In the UK the total number of holdings with pigs in 1991 was 17,226 and of these 12,429 were breeders with sows and gilts. By assumption, the difference of 4,797 must have been holdings with feeding-only pigs. If one also assumes that a similar number of the breeders sold the progeny of their herds as weaners, then systems of pig production can be estimated as shown in Table 2.2 for the UK and for England and Wales.

Table 2.2 Estimated systems of pig production 1991

Type	England & Wales	United Kingdom
	%	%
Breeding & feeding	40	44
Mainly breeding	30	28
Feeding-only	30	28
	100	100

Source: Agricultural Census, MAFF

Larger herds

The frequency distribution tables of census results provide an indication of the proportion of the pig population in the larger herds. For this purpose larger herds are defined as those with 200 and over sows and gilts for breeders and those with 1,000 and over pigs of all types for total pigs.

In the UK, 1,030 holdings (8 per cent of the total number of breeder herds) had 200 or more sows and gilts. Between them they had 54 per cent of all sows and gilts. The North Eastern, East Midlands, Eastern and Southern regions and Scotland had high proportions of their total sows and gilts in the larger herds of 200 or more sows and gilts, while the Northern, South Eastern and North Western regions, Wales and Northern Ireland had smaller proportions.

The UK had 2,167 holdings with 1,000 or more total pigs (13 per cent) and 71 per cent of all pigs. The same regions had the highest proportions of their pigs in the larger herds as for breeders. The dispersion around the UK average was less for total pigs, with only Wales and Northern Ireland having a smaller percentage of their pigs in larger herds. The actions of the few large producers and the subsequent effect on demand and prices will have an over-riding influence on what happens to UK pig production in the future.

Table 2.3 Larger herds by region 1991

Region	Breeders				Total			
	Holdings		Sows & gilts		Holdings		Total pigs	
	No	%	No	%	No.	%	No.	%
Northern	18	6	7,381	44	44	10	96,426	65
N. Eastern	266	15	103,946	59	521	22	1,285,196	76
E. Midlands	88	11	39,842	57	175	16	507,119	74
Eastern	218	15	93,116	58	476	23	1,161,849	76
S. Eastern	17	4	7,266	45	49	8	114,165	70
Southern	147	11	62,906	63	256	14	669,980	76
S. Western	78	5	32,798	51	161	7	441,959	70
N. Western	74	6	23,167	40	182	11	365,771	63
England	906	11	370,422	56	1,864	15	4,642,465	73
Wales	8	1	2,787	24	18	1	45,663	45
Eng & Wales	914	10	373,209	56	1,882	14	4,688,128	72
Scotland	70	13	34,293	67	132	15	383,735	78
N. Ireland	46	2	15,090	26	153	6	322,316	55
U.K.	1,030	8	422,592	54	2,167	13	5,394,179	71

Source: Agricultural Census, MAFF

The Cambridge region

The Eastern region, as designated for the University agricultural provinces, was the area covered by the Cambridge Pig Management Scheme. It included the counties of Norfolk, Suffolk, Essex, what was the Holland division of Lincolnshire, Cambridgeshire, Bedfordshire, Hertfordshire and part of Greater London. In 1991, the region contained 20 per cent of the breeding sows and gilts in the UK (24 per cent of those in England and Wales) and 21 per cent of total pigs (25 per cent England and Wales). With this large national share, the region is an important one for pig production and is only matched by the North Eastern region, which is another area traditionally associated with pigs.

Of the larger herds, 15 per cent of all holdings with sows in the Eastern region had more than 200 sows and together these units kept 58 per cent of all the sows in the region. The other 85 per cent of holdings had just 42 per cent of the sows in the region. For total pigs, the larger herds with over 1,000 pigs formed 23 per cent of all holdings with pigs in the region, and between them they had 76 per cent of the region's pigs.

Long-term changes

Frequency distribution tables of the census results show that the number of holdings with pigs has fallen drastically since figures were first published for England and Wales in 1955. Then the total number was 149,241, with 93.5 per cent of them small herds of less than 100 pigs in total. Only 77 (0.05 per cent) had more than 1,000 pigs. By 1992, the total number left in pigs was down to 13,279, a 90 per cent decline in 37 years, though this is probably an over-statement caused by changes in census definitions and method of counting holdings during intermediate years. Even so, these figures clearly illustrate the rapid exodus out of pigs.

With financial pressures mounting almost continuously on many pig producers, a pattern of a few pigs on every farm was not sustainable. Production efficiency was low, as many were making do with unskilled labour and unsuitable pig housing. On average, output in the late 1950s and early 1960s was about 14 pigs per sow in herd a year and 6.5 kg of feed, the major cost item, was required to produce 1 kg deadweight of pigmeat. (Corresponding performance for the early 1990s had improved to 21 pigs per sow and a feed requirement of 4.1 kg.) As there was a huge variation of individual results around these average figures, many producers failed to achieve even these levels of performance. For many of these, continuing to produce pigs was not a feasible proposition.

The producers that managed to achieve reasonable results stayed in pigs, at least for a time. Some specialised in pigs and with considerable capital investment expanded their units, a few producers did so almost continually. The 1992 census showed that the 14.3 per cent of the holdings in England and Wales with 1,000 or more pigs had 73 per cent of the total number of pigs. At the other end of the scale, the 56 per cent of holdings with small units of less than 100 pigs had 2.2 per cent of the national herd.

Table 2.4 Distribution of holdings and total pigs by size groups
(England and Wales)

	1-99	100-999	1,000+	Total No of holdings
Holdings	%	%	%	'000
1955	93.56	6.39	0.05	149.2
1960	91.1	8.8	0.1	110.7
1965	80.0	19.6	0.4	94.6
1970	74.9	23.9	1.2	59.9
1975	65.5	30.7	3.8	33.3
1980	60.0	32.8	7.2	23.0
1985	57.5	32.2	10.3	18.0
1990	56.0	30.7	13.3	13.6
1991	56.6	29.7	13.7	13.7
1992	56.8	28.9	14.3	13.3

	1-99	100-999	1,000+	Total No of pigs	Total pigs per holding
Pigs	%	%	%	'000	No
1957	52.6	45.0	2.4	4,759	34
1960	46.6	48.9	4.5	4,337	39
1965	30.6	59.7	9.7	6,178	65
1970	18.6	62.3	19.1	6,408	107
1975	9.1	52.3	38.6	6,337	190
1980	5.1	40.4	54.5	6,608	288
1985	3.4	32.4	64.2	6,792	378
1990	2.5	26.5	71.0	6,391	470
1991	2.4	25.5	72.1	6,497	473
1992	2.2	24.3	73.5	6,495	489

Source: Agricultural Census, MAFF.

The distribution of pig numbers by size groups was first published for 1957, two years later than holding numbers. Some 52 per cent of the pigs were then in small herds of less than 100 total pigs and just over 2 per cent in large herds of more than 1,000. With the decline in the number of producers, the average number of pigs per holding has increased steadily throughout, from 34 in 1957 to 489 in 1992.

Over the years, the structure of pig production has been reshaped from one where pigs were kept in relatively small numbers on many farms, to one where the vast majority of pigs are kept on far fewer farms but in much larger numbers. The future may see a continuation of this trend, though at a lower rate than previously. Many of the small units are on small family farms where usually they provide an important contribution to income, especially when they are housed in buildings that would otherwise be left empty, and are attended by family labour. Providing the returns from sales can cover the variable costs, many of these small herds are likely to remain.

Increased herd size

The increase in the average size of pig herds nationally has been reflected in the sample of the Cambridge Pig Management Scheme as shown in Table 2.5. The number of pigs produced per herd includes all sales, except cull boars and sows, and includes any home-bred gilts reared in the herd and retained for breeding stock replacements. The value of output has been calculated as pig sales less pigs purchased, plus or minus the difference between the opening and closing valuations.

Between 1955 and 1990 the average number of pigs produced per herd in the scheme increased sevenfold from 484 to 3,361. The value of output per herd rose even more, mainly due to inflation, by over twentyfold from £8,450 to £183,439. The profit margin in current terms for the much larger herds in 1990 was on average 24 times greater than that of 1955 but in real terms it was only just over double. Pigs were profitable in 1990 with output and margins correspondingly high. By comparison, 1991 was a poor year for pigs and in real terms the margin per herd was only 16 per cent of that for 1955, despite seven times as many pigs being produced per herd.

Table 2.5 Changes in average herd size and output

Year	Number produced per herd	Value of output per herd	Margin per herd[a] Current terms	Real terms[b]
		£	£	£
1955	484	8,450	1,325	15,635
1960	642	9,129	1,283	13,457
1965	924	12,577	1,398	12,568
1970	1,274	18,298	2,976	21,610
1975	1,650	50,673	11,184	45,760
1980	2,768	110,331	16,691	34,199
1985	3,037	138,962	14,508	20,550
1990	3,361	183,439	31,920	34,248
1991	3,408	154,847	2,457	2,457

Source: Cambridge PMS.

(a) Before interest charges
(b) Reflated by the Retail Prices Index at 1991 values

Chapter 3

The Changing Industry

The changing regimes

During the 1939-45 war the pig population in the UK was allowed to run down, as priority for the use of scarce feedingstuffs went to milk production. Since then there have been three distinct phases of pig production.

1. Government control - 1945 to 1954
2. Free trade within the UK with guaranteed prices - 1954 to 1973
3. Membership of the European Community and the subsequent effects of the Common Agricultural Policy, from 1973

The introduction of a single market for EC countries in 1993 may prove to be the fourth phase. Freedom to trade across national borders of Member States could increase the movement of pigmeat throughout Europe, affecting supplies both in and out of the UK, which in turn will influence pig prices and the level of home production.

The first phase after World War II was one aimed at recovery, to re-establish a neglected pig industry. In 1947, the sow population in the UK was down to 200 thousand, compared to 550 thousand pre-war. Annual slaughterings of finished pigs (excluding cull sows and boars) were only 1.6 million compared to over 6 million. All meat was in very short supply and severely rationed. Pigs for slaughter were strictly controlled and were bought by the Ministry of Food. Government policy then deliberately raised pig prices to producers to stimulate expansion and allow the abolition of meat rationing. All farmers were encouraged to keep pigs, despite the fact that feedingstuffs were still rationed and remained so until 1953. In 1950, for example, the flat rate price paid for pigs between 63 and 82 kg deadweight, then seven to nine score (140 to 180 lb), was equivalent to 27p per kg (49s 6d per score). In real terms, at the 1992 value of the pound, this was equal to 390p per kg, an incredible price by any modern standards. Feed then averaged £24 per tonne (2.4p per kg), equal to a feed : pig price ratio of 11 : 1, and despite poor standards of production, margins were naturally very good and at a record level of £30 per £100 output the best herds did even better. To survive in pig production through to 1992, when feed averaged £161 per tonne and pigs made 114p per kg (a ratio of 7 : 1), suggests that substantial improvements in performance have been achieved during the intermediate years.

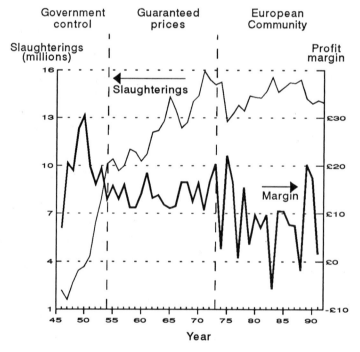

Figure 3.1 Annual pig slaughterings and average profit margin per £100 output

18

By 1954, the year of decontrol, pig production had increased sixfold from 1947 to 10 million slaughterings a year, clear evidence of the success of the campaign. The expansion occurred in time of considerable shortage when consumers were eager to buy more meat. It also shows how farmers can react strongly to favourable conditions.

Having achieved the desired results and with production then at a level sufficient to meet demand, free trade in pigs for slaughter was reintroduced on 1st July 1954, with a guaranteed scheme of government support aimed at providing reasonable profits for producers. This ensured minimum prices for pigs, which if not on average met by the market were made up in the form of deficiency payments. The guaranteed price was fixed at the annual review in relation to a given level of production and adjusted, up or down, according to actual slaughterings and for changes in feed prices as measured by the cost of a standard ration. The 18 years of the guaranteed scheme were, with hindsight, a good time for many pig producers, especially in view of what was to follow.

Entry to the EC in 1973 brought an end to the guaranteed scheme and started the phase of pig production governed by the CAP. Unlike beef and sheepmeat, pigmeat received no direct support in the form of intervention purchases or deficiency payments - only protection from imports from countries outside the EC and occasional aid for private storage in times of over-production, though there has been little uptake in the UK. In addition, there was a highly complex arrangement of monetary compensatory amounts (MCAs) designed to mitigate the disturbances for intra-Community trade which could arise from changes in exchange rates. These subsidies or levies were intended to leave the return to the exporter in his own currency as it would have been at the previous exchange rate. Many in the UK pig industry, however, were of the opinion that these regulations did at times distort trade and producer prices.

The effect of the Common Agricultural Policy

The EC pigmeat regime had been introduced some six years before the UK entry. The difference then in price levels between the EC and the UK for both meat and cereals was substantial and to help the changeover, the UK was granted a seven year transitional period. During this time pig prices increased from 30p to 78p per kg deadweight (157 per cent) and feed prices rose even more from £35 to £117 per tonne (233 per cent). Labour and other costs were also rising rapidly, mainly due to high inflation at that time. Production costs were therefore rising at a faster rate than finished pig prices, where the increases were meeting some consumer resistance. Without the steady improvement being achieved in efficiency, producers would have been considerably worse off. Much of the rise in average performance could be attributed to better breeding stock, quality of feed and other technical improvements, but some was due to the large reduction in the number of producers. Between 1970 and 1980 some 60 per cent of producers went out of pigs. Usually it was the less efficient that gave up and their departure helped to raise the averages of those that remained.

Joining the EC brought an end to any direct exchequer payment for pigs and since then producers have been exposed to the full force of market fluctuations in prices they receive for their pigs. As a result profit margins have varied considerably. Higher prices for beef and lamb could increase the demand for pigmeat but many felt aggrieved that unsupported pigmeat was having to compete with often heavily subsidised beef and lamb. In 1990-91, for example, only £6 million was spent in the UK on pigmeat, whereas £437 million was paid for beef, £328 million for sheepmeat and a further £263 million went for milk products.

Pigmeat supplies

The 14 million slaughter pigs produced each year in the early 1990s provided some 950,000 tonnes of pigmeat, worth about £1 billion. In addition, just over 300,000 tonnes were imported, three-quarters as bacon. Denmark, Holland and Ireland were the main suppliers, with smaller quantities from France and Germany. Some UK pork (including pigmeat from sows) has been exported, - the quantities involved have usually been offset by imports of pork. A small quantity of about 5,000 tonnes of bacon a year has been exported - a tiny amount when compared with imports (Table 3.1).

Table 3.1 Pork and bacon supplies in the UK ('000 tonnes)

	Home-produced	Imports	Exports	Net supplies	% Home-produced
1970-75 Av.					
Pork	645	21	11	655	98
Bacon	253	336	1	588	43
1975-80 Av.					
Pork	637	27	14	650	98
Bacon	214	293	2	505	42
1980-85 Av.					
Pork	718	34	30	722	99
Bacon	205	282	6	481	43
1986 Pork	748	32	51	728	103
Bacon	206	258	6	459	45
1987 Pork	771	46	44	772	100
Bacon	195	260	5	451	43
1988 Pork	799	57	53	803	100
Bacon	199	256	5	450	44
1989 Pork	725	88	52	759	96
Bacon	193	260	5	448	43
1990 Pork	740	77	49	769	96
Bacon	179	260	5	434	41
1991 Pork	778	71	73	775	100
Bacon	175	254	5	424	41
1992 Pork	780	80	91	771	101
Bacon	169	234	5	397	43

Source: MAFF, MLC

During the 1980s total net supplies of pork available for consumption increased to 800,000 tonnes a year; more recently they appear to have stabilised at a slightly lower level, as shown in Table 3.1. Bacon supplies, on the other hand, have declined almost continuously as consumer demand has fallen. The UK is virtually self-sufficient in pork but only produces just over 40 per cent of its bacon requirements. This market share has existed for many years and has changed little, despite lower total supplies of bacon. It has been in the bacon sector that opportunities have existed for UK producers to increase their share of the market. Further penetration has proved difficult as both Denmark and Holland have been well established as suppliers to this country for a very long time, and have acquired a high reputation for their products. Although improvements in the quality of British bacon have led to claims that it is now as good as the best imported bacon, there has been no increased demand for it, but these improvements may have been instrumental in retaining its market share. Joining the EC in 1973 has had little effect on UK market shares of pigmeat. The percentage of total supplies formed by home production has remained fairly constant throughout.

It remains to be seen whether the pigmeat trade between EC countries changes as a result of the Single European Market introduced in 1993. In view of previous experience the impact on UK pigmeat supplies may not be too drastic. Perhaps more pigmeat products, such as sausages and salami, will arrive to tempt UK consumers. At the same time, further opportunities will emerge for UK exporters to expand, especially with specific cuts or products tailored to meet the requirements of the importing countries. Some anxieties exist about implementing EC legislation fairly across all member countries, which otherwise could disadvantage UK production. A number of sensitive issues concerning animal health and welfare, the environment and food safety, need to be considered and the same enforcement applied evenly to all member countries. Public awareness about these matters could be beneficial to the UK industry in retaining or increasing consumer demand for home-produced pigmeat.

Competition from other meat

As already shown in Table 3.1, home-produced pigmeat competes with imported pigmeat and in addition there is competition from other meat. Poultry has overtaken beef as the most popular meat; though beef is still first choice for many people, chicken has gained popularity on price.

Table 3.2 Consumption of meat in the UK
('000 tonnes)

	1980-85	1987	1989	1990	1991	1992
Beef and veal	1,101	1,153	1,063	1,003	1,002	1,001
Mutton and lamb	412	376	411	429	424	378
Pork	722	772	759	769	775	771
Bacon	481	451	448	434	424	397
Poultry	825	1,017	1,061	1,108	1,133	1,192
Offal	268	247	233	219	210	208
All meat	3,808	4,015	3,974	3,962	3,967	3,947
	%	%	%	%	%	%
Beef and veal	28.9	28.7	26.7	25.3	25.2	25.4
Mutton and lamb	10.8	9.4	10.3	10.8	10.7	9.6
Pork	19.0	19.2	19.1	19.4	19.5	19.5
Bacon	12.6	11.2	11.3	11.0	10.7	10.0
Poultry	21.7	25.3	26.7	28.0	28.6	30.2
Offal	7.0	6.2	5.9	5.5	5.3	5.3
All meat	100	100	100	100	100	100

Source: MLC

Pork and poultry have increased their share of all meat consumption. The rest have declined by varying amounts. In 1992, more poultry was eaten than any other meat, 30.2 per cent of the total, followed by beef (25.4 per cent), and then pork (19.5 per cent). The decline in the amount of bacon consumed more than offset the increase in pork, so that total pigmeat now has a smaller share of all meat than before.

Chapter 4

Methodology

Purpose

The methods used in the Cambridge Pig Management Scheme were designed to measure as accurately as practicable the output, costs and standards of performance achieved in pig production. They were intended to provide not only valid comparisons between individual herds but also between groups of herds representing different systems and types of production.

Records required

The data required consisted of a monthly record of changes in pig numbers and feed used. These records were completed by the co-operating farmers and sent in for processing. A valuation of all pigs on hand was required at the beginning and end of each period. Labour and other costs were collected at the end of each recording period. Specially prepared forms were provided (see Appendix 5). All records were scrutinised for feasibility and where necessary checked against payment certificates and invoices. Experience showed that more confidence could be placed on records when they were collected monthly, as details were usually more readily available to check prices and any queries or discrepancies could be resolved promptly while details were fresh in mind.

Recording periods

The recording year ran from 1st October to 30th September and was divided into two periods, a winter half (October to March) and a summer half (April to September). With frequently fluctuating prices it was essential that all herds should be recorded over the same period of time.

Classification of stock

Details of breeding stock and feeding stock, together with their feed, labour and other costs, were recorded separately to permit the calculation of important efficiency factors and to distinguish between the two in assessing production costs. Breeding stock included all sows, served gilts and service boars. Young pigs, whatever age they were weaned, were included with breeding stock until eight weeks old, then transferred to feeding stock. Feed for the young pigs (piglets) was recorded separately. Transfer at a standard eight weeks was beneficial in regulating the calculation of weaner costs, comparing results and overcoming distortions in total costs assessments due to varying ages at transfer if related to weaning. Feeding stock included all pigs over eight weeks of age which were not being used for breeding. In larger herds, where the on-farm situation sometimes makes it difficult to distinguish between pigs under and over eight weeks of age, the following simple method was adequate for this purpose and often proved helpful.

To calculate the number of young pigs under eight weeks, take the births during the last two months (a month being approximately four weeks) and subtract the number of deaths in the same period. If greater accuracy is required, an allowance can be made for any two months that total nearer nine than eight weeks. All growing pigs above this number should then be classed as feeders over eight weeks of age.

Home-reared gilts and young boars required for breeding were recorded with feeding stock until they reached approximately 90 kg liveweight and were then transferred back to breeding stock. Purchased gilts and young boars were included with breeding stock.

When maiden (unserved) gilts were included with breeding stock their number was recorded separately from the number of sows and in-pig (served) gilts until they reached the time of first service.

Pigs purchased and sold were classified as follows:

Weaners	- under 30 kg liveweight
Stores	- over 30 kg liveweight
Porkers	- usually 40-60 kg deadweight, 55-80 kg liveweight
Cutters	- usually 60-75 kg deadweight, 80-100 kg liveweight
Baconers	- usually 65-75 kg deadweight - sold as bacon pigs
Heavies	- usually 75-95 kg deadweight, 100-120 kg liveweight
Feeder culls	- poor pigs sold for which payment was received (totally condemned pigs were classed as deaths).

In later years, piglets (under 15 kg) were occasionally traded and usually sold direct from the breeding herd, not having reached the eight week stage.

Valuations

A valuation of the pigs by category, with the liveweight of feeding stock and any feed on hand, was completed every six months at 1st October and 1st April. The numbers of pigs on each valuation had to agree with those on the monthly return for the same date. The closing valuation for one period also served as the opening valuation for the next period. In all cases values were assessed at Cambridge on the following basis.

Purchased boars, sows and gilts were depreciated from cost to cull value over their expected life in the herd. For simplicity, breeding stock purchased in the preceding six months were valued at cost at the next half-yearly valuation and others at a standard approximately half-way between cost and cull value. Home-produced unserved gilts retained for breeding were valued at slightly over bacon pig price; when served they were valued half-way between this and the standard rate, then at the standard rate when they became sows. The standard value was constant for both opening and closing valuations. When the need arose to increase the standard value, the opening valuation for the following year was recalculated at the new rate. High priced boars and grandparent stock were written down to the standard value over two years to avoid excessive depreciation in the first period.

Pigs under eight weeks of age were valued by weight or age according to the prevailing market value of weaners. Most co-operators found it easy to record the pigs in this category in two groups: unweaned pigs still on the sows and the remainder under eight weeks as weaned pigs. Each group could then be valued at an average amount scaled down from known market values.

Feeding stock were valued by weight on a scale assessed from the current market value of weaners and end product (porkers, baconers, etc.). To assess liveweights of feeding stock, co-operators were encouraged to weigh representative pigs from each pen or batch. Occasionally this proved impracticable because of the size of the enterprise or a reluctance to disturb the pigs. If measured weights were not available they were estimated. For herds with a regular throughput the following formula was helpful in estimating the average liveweight:

$$\frac{\text{Av. livewt brought in} + \text{Av. livewt sold in}}{2} \times 90 \text{ per cent*}$$

* For herds producing cutters or baconers, 95 per cent for lighter pork production

Using 90 per cent (95 per cent for lighter sales) of the average weight allowed for faster growth rates during the second half of the rearing period.

Any feed on hand which had been recorded as issued to pigs but not consumed by the date of valuation was also noted. This feed was deducted from the amount recorded for the current period and carried forward to the next.

Monthly records

The following details were recorded each month for stock and feed used.

For pigs purchased or sold numbers, prices and weights were recorded by category (weights were not required for breeding stock, apart from transfers from feeding stock). Any additional payments for pigs sold previously were written in separately with a note of which pigs they applied to. The costs of haulage of pigs, marketing charges, levies, etc., but not VAT (which was usually recovered), were added to the price of purchases and deducted from sales. If this information was not available when the monthly record was filled in, it was then collected at the end of the period. Deadweights of all pigs sold for slaughter were available from the payment certificates and these were later converted to liveweight (for the liveweight gain calculation refer to the formula or table in Appendix 3). Any contract level delivery bonuses were recorded separately and included as such in re ceipts for the year in which they were received, even though part may have applied to pigs sold previously. Any contributions received or paid for boars purchased or sold by buyers or sellers of weaners and stores were added to the value of weaners/stores purchased or sold. The value of boars was shown at their full cost.

Other information required for the monthly return included the total number of pigs born alive (but not born dead) and the number of litters farrowed, the age at weaning and average liveweight of young pigs at eight weeks of age. For pigs born, the number alive when the litter was first seen was recorded. Also recorded were the number and liveweight of any home-bred boars and gilts transferred from feeding stock to breeding stock. If any pigs purchased, sold or transferred had not been weighed a weight was estimated. Finally, the inclusion of the number of pigs on hand by category at the beginning and end of each month provided a check that all pigs had been accounted for before the return was sent in for processing.

The quantities and values (per tonne or total) of feed deliveries (compounds) or prepared (own mixed) were recorded separately for breeding stock, piglets and feeding stock. When the same feed was given to young pigs before and after eight weeks of age, all was often recorded to either piglets or feeding stock (whichever received the most) and then allocated between the two on a suitable feed quantity by age scale at the end of each six month period. All purchased compound and straight feeds were charged at the delivered costs to the farm, net of any discount received. If feed was collected by the farm lorry a charge for transport was added to feed costs, either monthly or at the end of the period. The quantity and cost of any growth promoters and additives which were incorporated with feed were included with feed (not veterinary costs) even if used for medicinal purposes. Home-grown feeds were valued at estimated market (farm-gate) price in the month of use. For cereals the values net of any co-responsibility levy (when applicable) were assessed at Cambridge from Home-Grown Cereals Authority data.

Feed mixed on the farm

To ensure greater accuracy and establish the types of feed used, own mixed feed was recorded and costed by each ingredient separately, usually on a per tonne, or per mix, basis. The costs of milling and mixing (and pelleting where applicable) were added to feed costs at the end of each period. When undertaken by a contractor, his charge was used. When undertaken by the farmer the charge included depreciation (10 per cent) of the current value of equipment and buildings (including storage), Pharmaceutical Society registration fee, repairs, replacements, running costs, labour and an allowance for milling loss. These were charged according to each farm's own costs and not at a standard figure. The cost of transporting prepared feed from the mill to the pigs was excluded here but included with farm transport in other costs.

Labour

Details of labour requirements were collected on a farm visit undertaken at the end of each recording period. Labour included the gross pay of pig managers, pigmen and women, any overtime and holiday pay, bonuses, relief for holidays, employer's share of national insurance and any pension contribution, the value of free housing, council tax payments and other perquisites. Also included was the cost of any other farm labour for work in connection with pigs, together with any contract, casual and secretarial work which could be reasonably allocated to the pig enterprise. Labour used for milling and mixing feed, for alterations and repairs to the buildings, haulage of pigs to or from the farm, and baling straw, was not included here but charged under the appropriate heading. Removal of manure from the piggeries was included but labour used for spreading manure on the fields was not charged to the pigs.

Any work on pigs undertaken by the farmer and his family was assessed by the number of hours worked. The farmer's own labour was charged at Grade 1 rates and family labour at the equivalent employed rate by age according to the type of work involved.

Other costs

As the heading suggests other costs include all other expenditure incurred by keeping pigs. These costs varied considerably between farms and in the early 1990s ranged from 10 to 25 per cent of total costs, depending on type of production and individual circumstances. It was, therefore, important to deal with these costs carefully in consultation with the farmer concerned, preferably during the half-yearly visit. By necessity some items were estimates but efforts were made to neither understate nor overstate these costs.

(a) Farm transport

The main charge was for the use of tractors and loaders but included motor journeys made specifically for pigs and movement of pigs on the farm. Haulage of pigs to or from the farm and of feed from the merchant to the farm was not included here, but charged to those items. If the pig enterprise had exclusive use of a tractor the charge made covered depreciation (12.5 per cent per annum of current value), repairs and maintenance, fuel and oil, tax and insurance. Farm tractors and loaders used occasionally were charged at a standard rate per hour according to age and size. Tractor drivers were included under labour.

(b) Veterinary

This included the veterinary surgeons' fees and the cost of veterinary supplies (e.g. injections, medicines, disinfectants, etc.) but not feed additives and supplements, which were added to feed costs.

(c) A.I. fees

This included costs incurred for artificial insemination and equipment.

(d) Power and water

Costs for electricity and gas included heating, lighting, fans, feeding and cleaning equipment. Electricity for milling and mixing was added to feed costs. Water charges included that used for drinking and cleaning.

(e) Miscellaneous costs

Included small tools, replacement heaters, lamps and bulbs, markets, ear tags, protective clothing, subscriptions, recording fees, weighbridge fees, computer fees, consultancy fees, accountant's fees, insurance premiums, telephone, stationery, postage, pest control, etc.

(f) Litter

Purchased straw, wood shavings and other litter were charged at cost and a deduction made for pig manure when it was of value to the farm. Home-grown or free straw was not valued (this assumes it equalled the value of manure) but costs of baling and carting (by contractor or own resources) were charged, unless a "straw for muck" arrangement existed with another farmer.

(g) Maintenance

The repairs and maintenance of pig buildings and equipment were included together with upkeep costs such as painting, creosoting and lime-washing. The costs of structural alterations were not included here but added to the buildings account.

(h) Buildings

A charge was made for all buildings used for pigs, even if they had been written off in the farm accounts. The buildings included straw shelters, bulk meal bins (not grain), feed store, workshop, office, staff rooms, loading bays, slurry storage, drainage, permanent equipment usually part of a building (feeding systems, etc.), concrete pads and roadways, etc. The annual charge was 10 per cent of the assessed total current value. New buildings were valued at cost. Other buildings were included at their depreciated value, reflated by the MAFF Index of Prices for Farm Buildings to represent current values. The updated values were then adjusted where necessary to take account of the present condition of the buildings and the costs of any major alterations and conversions. For buildings rented off the farm for pigs the actual rents paid were charged.

Building values depended on the selective judgment of the assessor. The aim was for these values to represent, as far as possible, realistic estimates relative to pig buildings on other farms in the scheme. These values were adjusted by such factors as general impressions, aspect, services provided (roadways, drainage, etc.), construction and condition of floors, walls, roofs, partitions and contents.

(i) Equipment

This usually consisted of feeders, drinkers, heaters, weighers, trolleys, trailers, muck-scrapers, pressure washers, slurry tankers, service and farrowing crates, fencing, computers and office equipment, etc. All items of equipment in use were included and their opening value assessed by the diminishing balance method at a rate of 20 per cent per annum. The cost of new equipment purchased during the period was added to the opening valuation and the total charged at 20 per cent per annum and the balance carried forward to the next period. Any charges for hire of equipment were also included.

(j) Pasture

Pasture was charged for at the current rent per hectare for equivalent land in the district, plus a charge to cover cost of cultivations, seed, etc. If the pigs did not have the exclusive use of the land only a proportion was charged.

Costs omitted

None of the following items were included in the costs of production for individual herds because they usually applied to the whole farm and were difficult to allocate for the share to pigs:

1. Interest on capital invested
2. Bank charges
3. Overhead expenses for the whole farm unless expressly stated

Presentation of herd results

Every co-operator in the scheme received their own results[1] promptly at the end of the six month recording period, usually within a fortnight of the completion of the herd records. Average results of all herds, for comparative purposes, followed later when all records had been finalised. Each year, the two six-monthly results were combined to give the annual results.

The summary of results (see Appendix 5) sent to members was on a net margin basis and presented in three parts:

1. A trading account of financial results
2. Quantitative details of performance achieved
3. Average weights and prices of pigs purchased and sold

To allow comparisons of herds with different systems, different types of production and varying sizes to be undertaken, the financial results of the trading account were summarised on the basis of £100 of livestock output. Details of the calculation of livestock output were also given and this, basically, was the difference in value between income from pig sales and expenditure on pig purchases, adjusted for changes in opening and closing valuations. The latter could be because of changing values or changing numbers.

$$\text{Livestock output} = (\text{pig sales} + \text{closing valuation}) - (\text{pigs purchased} + \text{opening valuation})$$

Feed, labour and other costs were related to output per £100. If the total of these costs was less than output, the resulting positive margin indicated a surplus (profit); if it exceeded output, the negative margin meant a loss had been incurred.

---

[1] The methods used to calculate results are given in Appendix 2.

Quantitative details were given separately for breeding stock and feeding stock for the key production factors having a major influence on profitability. From this list it was usually possible to demonstrate why a particular herd was more or less profitable than the average of others. Average weights and prices of pigs purchased and sold by category also provided the means to compare results with others in the scheme.

Comparisons with other schemes

Comparisons of results from the increasing number of recording schemes now on offer are fraught with danger because of varying methods and versions of dealing with fixed and variable costs. The three most common methods are as follows:

1. Margin over feed costs: the value of output less feed costs
2. Gross margins: the value of output less variable costs
3. Net margins the value of output less fixed and variable costs

Margin over feed costs is a basic method of measuring some financial results but provides little information to establish performance standards achieved. Feed costs are unlikely to include any charges for preparing feed milled and mixed on the farm. At best, it can be classed as a starter method for first-time recorders seeking experience.

With gross margins, it is necessary to distinguish between fixed and variable costs. Fixed costs are those not normally affected by small changes in the size of the enterprise, such as regular labour, equipment and buildings, machinery, power and water, accountant's fees, insurance, etc., while variable costs usually change proportionally with size and include feed, veterinary costs, etc. Here again, the cost of feed is unlikely to include milling and mixing charges.

While gross margin assessments are ideal for whole farm operations and many specific enterprise studies which have common use of land and machinery resources, they are often less satisfactory for specialised enterprises such as pigs. For example, a good gross margin on a pig unit does not necessarily signify a profitable herd. The advantage indicated may have been due to high investment in specialist buildings and skilled labour which, if charged, would largely offset the superior gross margin performance, whereas another herd with a lower gross margin may only have minimal other charges and finish up just as profitable.

For pig production the net margin approach is preferred. Problems of distinguishing between fixed and variable costs do not arise when all are included. It is important, however, to group these costs under appropriate headings; otherwise results can be misleading. Labour costs for milling and mixing, baling straw and maintenance work are best kept separately from labour for attending pigs. When combined, the results for a unit doing all these things with farm labour will look quite different from a unit using compound feed and contractors to undertake baling and maintenance. Unless separated, interpreting such results will take time and leave unclear how labour costs for this unit compare with others.

Using a lay-out similar to that used in the Cambridge scheme will allow straight-forward comparisons to be made. It then quickly becomes apparent why some herds are more profitable than others.

Chapter 5

Costs and Returns

Feed and pig price

The overall level of profitability of pig production is largely determined by the relationship between feed and pig prices. When the differential narrows, margins are squeezed and as it widens, they increase. Individual herds vary around this average level according to the standard of efficiency achieved. The best, with low feed requirements, make more profit, while the worst, using more feed, make less. Feed has always been the major item of costs and although its share has been declining since 1985, because feed prices stabilised while labour and other costs increased, it still formed on average 67 per cent of total costs in 1991. This ranges from 58 per cent for mainly breeding herds to 76 per cent for feeding-only units. Consequently, a change in feed prices without a corresponding change in pig prices will have an impact on profit margins.

Between 1951 and 1970 feed prices were nearly static at around the £30 per tonne and only really took off when the UK joined the EC in 1973. After almost continuous escalation feed peaked at £173 per tonne in early 1984, a rise of over 470 per cent in 14 years. In real terms (see Appendix 1), at the 1991 purchasing power of sterling, feed prices reached £400 per tonne at the beginning of 1974. In the following years monthly prices fluctuated between £240 and £350 per tonne and from 1984 to 1992 they fell almost continuously. These average prices covered both breeding and feeding stages of production to slaughter weight and included both purchased compounds and home-mixed feeds used by the herds in the scheme.

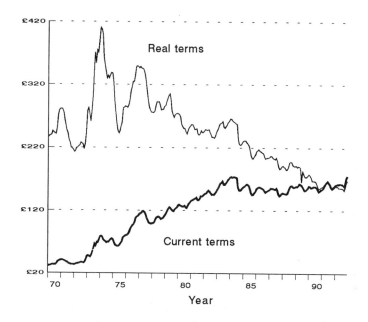

**Figure 5.1 Monthly average feed price per tonne
used by the herds in the scheme in current and real terms
(breeding and feeding)**

Since 1984 monthly feed prices have fluctuated seasonally
but in current terms they remained below the peak until the end
of 1992 when the green pound devaluations and the consequen-
tial increase in sterling intervention and market prices caused a
substantial rise.

Pig prices in current terms followed a similar trend to feed
until 1970, varying between 25p and 30p per kg deadweight. By
1984, they had reached 114p per kg, a rise of about 300 per cent -
only two-thirds of the increase in feed prices. Pig prices declined
steadily for the next three years and since then have fluctuated
quite dramatically, due primarily to short-term variation in
supply and demand for pigmeat. Peaks, as measured by the
average all pigs price (AAPP), were in October 1989 (133p),
June 1990 (138p - the record), March 1991 (113p) and April 1992
(128p). Troughs occurred in February 1988 (86p), January 1990
(102p), December 1990 (92p) and August 1991 (83p - the lowest
for 12 years).

37

In real terms pig prices were also much higher and they too peaked in 1973 and 1974 at around 270p per kg deadweight. Since then they have fallen steadily, despite a number of fluctuations on the way, to a little over 100p per kg at the end of 1992. Pig prices in 1992 (a more normal year than 1991) were some 55 per cent of those prevailing in real terms for the three years of 1970-72 and had fallen more than feed prices, which were 66 per cent of the 1970-72 base.

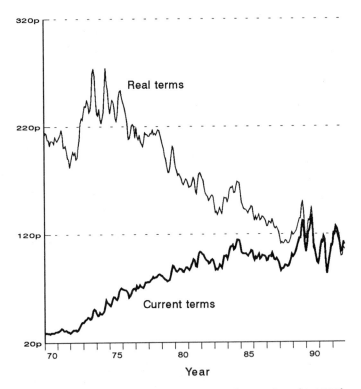

**Figure 5.2 Monthly average all pigs price (AAPP)
per kg deadweight in current and real terms**

Traditionally, pig prices changed seasonally, being at their best during the autumn in preparation for the Christmas trade. Usually they were weak for the first couple of months of the new year, then improved gradually in the spring. Summer prices normally reflected the weather, suffering when hot but holding up reasonably well in the cool seasons. More recently, seasonal trends in pig prices have disappeared and no regular pattern now exists. Both high and low prices have occurred at the beginning, middle and end of the year, regardless of weather conditions. In fact, in 1990 and 1991, the highest prices of the year were in mid-summer, something unheard of previously.

Movements in producer prices for pigs nowadays seem directly linked to the actions of a few large-scale buyers seeking supplies for the major supermarket chains. Most retailers aim for price stability to maintain the volume of business but in times of shortages, or anticipated shortages, they are prepared to pay extra to ensure future supplies. These wholesale prices soon fall, however, when supplies are plentiful, or when demand eases off. It is common these days for producer and wholesale price changes not to be fully passed on to consumers because of strong competition amongst retailers. Sometimes changes in retail prices bear little relation to producer prices. While no suggestion of collusion is intended, it is known that supermarkets do monitor each other's prices, as one would expect, and react accordingly. In reality, supermarkets, in setting their own competitive price levels, also determine producer prices, which are often in the short term unrelated to production costs. Since 1988 the prices paid to producers for their pigs have fluctuated more frequently and by greater amounts over longer periods than ever before.

Feed : pig price relationship

The old standard feed : pig price ratio, which related the cost of feed per kg to the value of pigs per kg to indicate the current viability of pig production, has now been superseded by the total feed cost as a percentage of the pig price. This method is more meaningful, especially for long-term comparisons, because it takes into account the quantity of feed used to produce a kg of pigmeat. As a result of substantial improvements, the quantity required now is far less than before (Figure 5.3).

The total quantity of feed used for breeding and feeding to produce 1 kg deadweight of pigmeat fell from 5.52 kg in 1970 to 4.10 kg in 1991, an improvement of 25 per cent. For the latter part of this period, most of this improvement came from the feeding stage of production, as breeding stock requirements had levelled off. As most herds were weaning early, further improvements in feed use were small, and these were offset by the increase in outdoor production, which used more sow feed than the indoor herds. The annual improvement in the feeding stage is now much less as efficiency has reached a high standard. It is comparatively easy to improve a poor performance but difficult to improve a good one, so chances of continually reducing overall feed requirements in the future are less likely. Occasionally there have been fractional set-backs, usually when health problems have been more prevalent, as in 1992 when porcine reproductive and respiratory syndrome (blue ear disease) infected several herds.

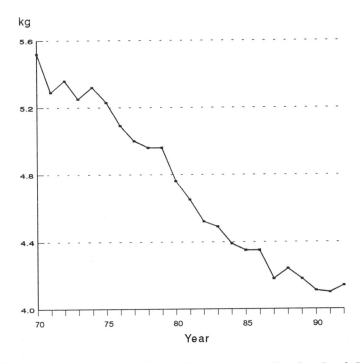

Figure 5.3 Total quantity of feed used per kg deadweight of pigmeat (breeding and feeding - annual average)

The three factors of feed price, feed quantity and pig price combine to give the total cost of feed required to produce 1 kg deadweight of pigmeat, as shown in Figure 5.4. The graph is on a monthly basis and the feed quantity has been seasonally adjusted to reflect the superior performance during the summer months, when more pigs are produced per sow in herd and less feed is used.

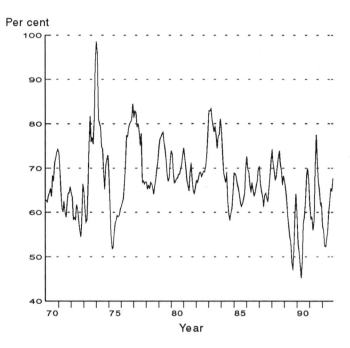

Figure 5.4 Total cost of feed as percentage of pig price
Monthly average feed cost for breeding and feeding to produce
1 kg deadweight of pigmeat (feed price per kg times quantity of
feed used - seasonally adjusted) as percentage of the pig price.

This graph (Figure 5.4) highlights clearly the fluctuating fortunes of pig producers. The worst time was in 1974, when for three months feed cost formed about 95 per cent of the pig price. This was a period of rapidly rising feed prices, while pig prices had slumped from over-production. Other bad times were in 1977 and 1983, when for more prolonged periods the cost of feed hovered around the 80 per cent mark. In 1977 escalating feed prices were again responsible and in 1983, a collapse in pig prices was caused by high slaughterings due to the Aujeszky's Disease Eradication Programme. More recently unprofitable times have been of relatively short duration with feed costs peaking at about 70 per cent of the pig price. By then, however, labour and other costs formed a much larger part of total costs, previously dominated by feed, and pushed pig production into the red. The best times were experienced when feed costs formed less than 60 per cent of the pig price. These were usually for shorter periods, the most recent being during the summers of 1989 and 1990, when pig prices were high and feed costs fell briefly to below 50 per cent of their level.

Prices in real terms

After two decades of relative stability during the 1950s and 1960s, prices increased drastically in the 1970s and early 1980s. Inflation, as measured by the retail prices index, increased at an even faster rate for several years, so that in real terms both feed and pig prices fell. Details of annual average prices are shown in Table 5.1.

Feed prices in real terms were at their highest in 1974, when they were 234 per cent above the 1992 average. They fluctuated for the next ten years, though on a downward trend, and they have fallen continuously since 1984. Pig prices were also highest in real terms in 1974. Here again pig prices have overall moved steadily lower despite partial recovery from time to time. For the three years 1990-92, feed prices averaged 68 per cent of their 1970-72 level, whereas pigs were lower still at only 56 per cent of this base. Without a steady improvement in production efficiency, shown earlier in Figure 5.3, in terms of the total quantity of feed used to produce one kilogram deadweight of pigmeat, pig producers would have been in serious financial difficulties for many years.

Table 5.1 Annual average feed and pig prices
in current and real terms[a]

Year[b]	Feed price[c] per tonne		Pig price[d] per kg dwt		Inflation[e] %
	Current	Real	Current	Real	
1970	32.08	242.26	28.66	216.43	5.8
71	38.79	269.26	30.86	214.21	8.8
72	35.22	227.26	30.70	198.10	7.6
73	49.63	295.29	40.36	240.14	8.4
74	72.23	376.04	47.96	249.68	14.3
75	70.56	300.25	58.50	248.93	22.3
76	85.40	304.87	67.12	239.61	19.2
77	110.88	340.17	70.82	217.27	16.4
78	104.02	291.77	78.19	219.32	9.4
79	117.22	296.16	79.88	201.82	11.0
80	126.71	270.01	87.34	186.11	18.6
81	135.77	256.67	90.64	171.35	12.7
82	146.60	251.86	98.48	169.19	10.0
83	157.35	257.77	91.37	149.68	4.9
84	168.87	263.45	104.32	162.75	5.0
85	155.67	229.32	105.46	155.35	5.9
86	152.21	215.67	99.94	141.60	4.0
87	153.10	208.63	98.45	134.16	4.0
88	150.13	196.16	90.97	118.86	4.3
89	158.51	192.59	106.54	129.45	7.5
90	156.85	175.01	119.35	133.17	8.9
91	158.59	164.93	101.22	105.27	7.3
92	160.80	160.80	113.90	113.90	4.0

(a) Reflated by the retail prices index to 1991 value
(b) Year ended 30th September
(c) Feed used by all herds in scheme
(d) AAPP (gross price before deducting marketing charges)
(e) As measured by the retail prices index

Drastic changes in feed prices are likely to occur in 1993. Firstly, the fall in the value of sterling in September 1992 caused imported protein feed prices to rise, and the subsequent green pound devaluations have raised cereal intervention prices. As intervention provides the floor in the cereals markets, the price of feed grains has also risen, as unchanged EC rates were converted to new sterling values. Secondly, the pressure on feed prices should be eased considerably from July 1993 onwards when cereal intervention prices will be reduced by 20 per cent as part of CAP reform measures. In addition, feed wheat will no longer be eligible for intervention. Average feed prices for 1993 should be high for the first six months and lower for the second six months. Both 1994 and 1995 should see still lower prices as two further steps of planned CAP reform take effect.

It is difficult to surmise what the level of pig feed prices will be in 1996 as a result of these proposed changes. In the past some expected changes failed to materialise because altered trading conditions prevented them. Obviously, much will depend on the value of sterling and its relationship with other currencies. Certainly feed prices are expected to fall and some projections of likely impact on costs of pig production are possible. At 1991 average efficiency and level of costs, a £10 per tonne reduction in feed prices would cut costs of breeding and feeding to bacon weight of 70 kg deadweight by £2.80 per pig (4p per kg). This is equal to £1 per pig for producing 30 kg weaners and £1.80 per pig for feeding-only herds. What happens to pig prices is even more fascinating. Will recent prices prevail and pig production become more profitable or will prices fall and the savings be passed to others? Probably some of each would be fair but many pig keepers will ask when fairness ever came into pig production.

Composition of costs

Feed still remains the major cost item but in recent years its importance has been marginally declining. While actual feed prices per tonne have been relatively stable, labour and other costs have been steadily rising and now form a far greater proportion of total costs than hitherto, as demonstrated in Table 5.2.

Labour and other costs, as a percentage of total costs, are highest for mainly breeding herds and lowest for feeding-only herds, with combined breeding and feeding herds roughly half-way between the two.

44

Table 5.2 Percentage composition of costs

		Breeding & feeding %	Mainly breeding %	Feeding only %
1970	Feed	77.0	66.9	85.6
	Labour	11.7	16.2	7.3
	Other costs	11.3	16.9	7.1
		100	100	100
1975	Feed	78.5	68.6	86.8
	Labour	11.3	16.2	7.0
	Other costs	10.2	15.0	6.2
		100	100	100
1980	Feed	76.0	65.3	83.8
	Labour	11.4	16.4	7.5
	Other costs	12.6	18.3	8.7
		100	100	100
1985	Feed	73.2	64.0	85.2
	Labour	13.0	17.6	6.7
	Other costs	13.8	18.4	8.1
		100	100	100
1990	Feed	69.0	59.2	76.9
	Labour	14.9	20.3	9.5
	Other costs	16.1	20.5	13.6
		100	100	100
1991	Feed	67.6	58.4	76.1
	Labour	15.6	21.1	10.3
	Other costs	16.8	20.5	13.6
		100	100	100

In 1975, labour and other costs together formed on average 21 per cent of total costs for breeding and feeding herds, but ranged from 31 per cent for mainly breeding herds selling weaners to 13 per cent for feeding-only herds. By 1991, the last year of the Cambridge scheme, labour and other costs had increased by over ten percentage points for each of the three systems to 32, 41 and 24 per cent respectively. This changing cost structure means that labour and other costs have become far more important and, most likely, will continue increasing at least until the mid-1990s as the reform of the CAP takes effect.

Livestock output

One of the most satisfactory methods of measuring and comparing costs and returns for individual herds, or groups of herds, run on a continuing basis, is in relation to the value of livestock output. This provides a standard way of comparing financial results for herds of differing sizes, varying systems and types of production. Livestock output is easy to calculate and has long been accepted as an equitable method of undertaking these comparisons. It is simply the value of pig sales added to the closing valuation of pigs on hand at the end of the period, less the cost of pigs purchased and the opening valuation of pigs at the start. Any increase in herd size is allowed for in the closing valuation and contributes to livestock output accordingly.

$$\text{Livestock output} = (\text{sales} + \text{closing valuation}) - (\text{purchases} + \text{opening valuation})$$

Costs of feed, labour and other costs are then deducted from output to leave the net margin. If output exceeds total costs, the margin is a surplus (profit) but if total costs exceed output, then the margin is a deficit (loss). A common basis for comparison is established when costs and margins are calculated for each £100 of output.

Output (all herds)

The most profitable years for pig producers were from 1947 to 1951 when herds were small and total UK production was only some 20 per cent of that for more recent years. From then until joining the EC in 1973, margins were more modest and fluctuated only marginally from year to year under the terms of the Guaranteed Scheme for pigs. The good and relatively stable times of this period quickly ended under the CAP and since then costs and margins have fluctuated more drastically, as shown in Figure 5.5. The worst years were 1983 and 1988, when the value of output failed to cover costs. The years of 1974, 1977 and 1991 were only a little better.

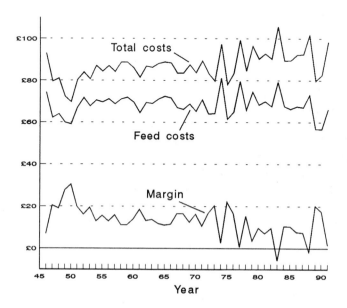

Figure 5.5 Average costs and margins per £100 livestock output (all herds)

Until 1972, £100 would buy five or six bacon pigs. Rapid inflation reduced this to 1.5 bacon pigs by 1982 and to 1.3 pigs by 1992.

Range of profitability

The margins given in Figure 5.5 are the averages for all herds in the scheme and averages usually conceal wide variations in individual results. The 20 most profitable and 20 least profitable herds in the Cambridge scheme have been selected to indicate the range between the best and worst herds and these are shown in Table 5.3.

The 20 most profitable herds managed to achieve on average margins of some £10 to £16 per £100 output more than the average of all herds in the scheme, while the 20 least profitable were between £14 and £26 per £100 output worse than average. In fact, the least profitable herds lost money on their pigs in 19 of the 22 years shown, often quite considerable sums. Comparing the most profitable with the least profitable, the difference in margins per £100 output ranged between £26 and £40.

Table 5.3 Range of margins per £100 livestock output

	Number of herds	All herds average	20 most profitable	20 least profitable
		£	£	£
1970	110	16.27	28.04	-5.59
1971	110	10.65	23.27	-12.11
1972	110	16.63	29.42	-5.49
1973	128	20.32	33.81	-2.82
1974	141	2.69	13.36	-18.74
1975	137	22.07	32.39	3.97
1976	148	16.61	32.38	-4.60
1977	146	0.71	13.17	-18.99
1978	146	15.37	27.99	1.51
1979	158	3.62	19.33	-22.43
1980	146	9.69	22.60	-15.04
1981	151	7.08	20.17	-12.36
1982	161	9.70	21.99	-7.55
1983	144	-5.70	8.94	-29.74
1984	151	10.49	22.45	-8.59
1985	150	10.44	20.91	-6.65
1986	144	7.67	17.73	-10.84
1987	148	7.48	19.80	-15.84
1988	147	-1.86	12.02	-24.22
1989	142	20.23	30.67	5.33
1990	152	17.40	31.02	-8.85
1991	150	1.59	13.41	-24.55

For the five years 1987-91, the 20 most profitable herds averaged a surplus margin of £21.38 per £100 output, while the 20 least profitable herds averaged a deficit margin of £13.62, amounting to an overall difference of £35 between the two groups. The value of finished pigs sold during this period averaged £65, which equals 1.54 pigs per £100 and, therefore, the £35 difference per £100 output was equivalent to £22.73 per pig. Such a significant difference in margins between the most and the least profitable herds seems quite remarkable. As Table 5.4

Table 5.4 Range of production costs and pig prices between the most profitable and least profitable herds

	Breeding costs per weaner[a]		Feeding costs per kg lwt gain		Net price pigs sold per kg dwt	
	Best[b]	Worst[c]	Best[b]	Worst[c]	Best[b]	Worst[c]
	£	£	p	p	p	p
1970	4.54	6.59	13.1	16.8	27.6	25.9
1971	5.24	7.56	14.5	18.4	29.8	28.0
1972	4.63	7.63	13.1	17.0	29.1	28.3
1973	5.75	8.66	17.8	25.6	39.2	39.3
1974	8.35	12.45	25.7	34.2	46.7	45.1
1975	8.62	13.07	25.9	33.7	57.6	54.5
1976	10.35	15.71	29.8	39.8	67.3	66.4
1977	12.57	18.44	37.9	50.7	70.5	71.0
1978	12.15	18.39	35.5	47.7	76.4	73.7
1979	13.71	20.78	39.6	53.4	77.9	77.6
1980	15.14	22.37	40.3	55.5	83.5	84.3
1981	16.53	24.33	42.4	57.2	88.0	88.6
1982	17.83	24.98	45.4	59.3	96.1	95.6
1983	18.36	28.01	47.6	62.6	89.1	89.4
1984	18.69	29.16	48.9	66.0	102.4	102.6
1985	18.67	26.27	45.8	60.4	102.5	102.0
1986	19.69	27.13	46.6	60.2	97.3	96.9
1987	19.49	26.60	44.3	60.3	95.8	95.3
1988	19.10	27.70	44.7	57.3	88.6	87.7
1989	20.06	28.70	45.2	60.0	103.7	103.5
1990	20.92	29.87	45.0	61.4	118.0	116.4
1991	23.01	32.40	45.8	64.2	97.5	97.7

(a) To eight weeks of age
(b) Average of 20 most profitable herds
(c) Average of 20 least profitable herds

shows, this was not just a one year occurrence; every year the margins were always substantially different. Reasons to account for the vast contrast are worth investigating. Comparing financial results achieved by the two groups is the first step.

Table 5.4 shows that for both breeding costs per weaner produced and feeding costs per kg liveweight gain the difference between the best and worst herds was considerable, but this was not so for the net price received per kg deadweight of pigs sold. In fact it was not uncommon for the least profitable herds to have as good or a higher price per kg deadweight than the most profitable herds. This clearly establishes that the variation in profit margins was due mainly to standards of production efficiency achieved and only marginally to price received for pigs sold.

During the last five years covered by this table, the most profitable group of herds received on average just 0.6p per kg deadweight (39p per finished pig) more than the least profitable group. Practically all of the difference was accounted for by the variation in performance standards in both breeding and feeding stages of production.

As shown in Table 5.5 the most profitable herds on average achieved superior results throughout. They produced more pigs per sow in herd a year from more frequent farrowings of larger litters, both born and reared, to achieve 22.6 weaners per sow compared with 18.9 for the least profitable herds. Less feed was used for sows, piglets and feeders, and it was cheaper for all types, especially for feeders, whose feed was mainly mixed on the farm. Combining both breeding and feeding, the most profitable herds required 3.74 kg of feed to produce one kilogram deadweight of pigmeat, while the least profitable used 4.66 kg.

Labour and other costs were lower per pig for the most profitable herds as they benefited from the extra pigs per sow to share these costs. Stock depreciation was also lower, mainly due to higher prices for cull sows. The lower mortality rate in feeding stock, combined with comparatively inexpensive weaners brought in, resulted in a mortality charge of under half that of the least profitable herds.

Total costs and returns per pig for the 20 most profitable and 20 least profitable herds, together with the difference between the two groups, are given in Table 5.6.

Table 5.5 Range of performance standards between the most profitable and least profitable herds five year average 1987-91

	20 most profitable	20 least profitable
Breeding		
Litters per sow in herd a year	2.34	2.18
Age at weaning (days)	26	30
Live pigs born per litter	11.0	10.3
Weaners per litter	9.7	8.6
Weaners per sow in herd a year	22.6	18.9
Sow feed used per sow a year	1.23 t	1.32 t
Sow feed per weaner[a]	54.4 kg	70.3 kg
Piglet feed per weaner[a]	17.5 kg	18.2 kg
Cost of sow meal per tonne	£142.28	£143.35
Cost of piglet meal per tonne[a]	£245.60	£267.95
Compounds as % of total feed	60 %	63 %
Costs per weaner[a]	£	£
Feed	12.03	14.93
Labour	4.01	6.29
Other costs	3.93	6.50
Stock depreciation	.55	1.33
Total breeding costs	20.52	29.05
Feeding		
Liveweight of pigs produced	85 kg	85 kg
Daily liveweight gain	.68 kg	.59 kg
Mortality percentage	1.9 %	2.5 %
Feed conversion rate	2.44	3.05
Cost of meal per tonne	£150.05	£158.07
Cost of feed per tonne[b]	£147.53	£156.34
Compounds as % of feed	18 %	73 %
Costs per kg liveweight gain	p	p
Feed	35.9	47.6
Labour	3.8	5.4
Other costs	4.6	6.2
Mortality charge	.7	1.4
Total feeding costs	45.0	60.6

(a) To eight weeks of age
(b) Includes other feeds (mainly by-products) converted to meal equivalent

**Table 5.6 Difference in costs and margin per pig, 1987-91
(breeding and feeding)**

	20 most profitable	20 least profitable	Difference
	£	£	£
Feed	37.12	48.31	11.19
Labour	6.72	10.20	3.48
Other costs	7.18	11.05	3.87
Stock depreciation	.55	1.33	.78
Total costs	51.57	70.89	19.32
Pig price	65.07	64.68	.39
Margin	+ 13.50	- 6.21	19.71

Of the £19.71 per pig difference between the two groups, £19.32 can be explained by production costs and 39p by finished pig prices. Prices received for finished pigs varied from herd to herd depending on quality and market outlet but on average the amount was quite small compared to the variation in production costs. While it is essential to produce pigs of good quality, to enhance the presentation and competitiveness of British pigmeat, improvements in production efficiency have been more rewarding to producers than improvements in the price of finished pigs. A comparison of performance standards between the best and worst herds clearly illustrates a vast difference and shows that a potential exists for many to make more money from their pigs. It is, of course, difficult for a producer with an already efficient herd to do better and his efforts should be directed to maintaining a position at the top.

Chapter 6

Breeding Herd Performance

Weaner numbers

In managing a breeding herd the aim should be to produce each year large litters of good quality weaners, at as low a cost per weaner as possible, that convert feed efficiently and grow quickly to slaughter weight without becoming excessively fat. By far the most important factor in the economics of breeding pigs is the number of weaners produced per sow in herd a year. This factor is the result of three others:

1. Frequency of farrowing
2. Number of pigs born alive per litter
3. Number of pigs weaned per litter

The farrowing frequency (litters per sow in herd a year) is the essential factor and this all starts, of course, by making sure that sows are served properly at the right time. The mortality rate for suckling pigs and weaners has been purposely omitted, though it can be derived from the difference between the number of pigs born per litter and the number of weaners per litter. In pig breeding considerable emphasis is often put on young pig mortality but, when taken in isolation, mortality numbers and especially mortality percentages can be misleading. For example, compare a litter of twelve pigs of which two die with a litter of ten and one death. The first has a mortality rate of 16 per cent while the second has one of only 10 per cent and on this measure would appear superior. The emphasis should be rather on the number that survive; then the litter with ten pigs reared is clearly performing better than the one with only nine. Losses through mortality are important, of course, but it is the survivors or number of weaners per litter that really counts.

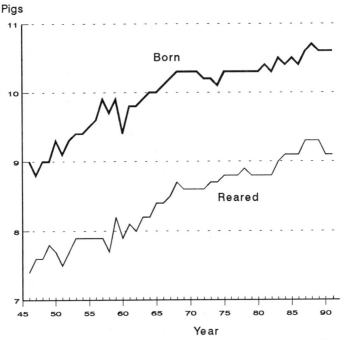

**Figure 6.1 Average number born and reared per litter
to eight weeks of age**

Litter size at birth and at eight weeks of age, shown in Figure 6.1, increased steadily from the late 1940s to 1987 and stayed at about this level until the scheme ended in 1991. From a base of 9.0 born and 7.6 reared per litter for the first five years, litter sizes rose to an average of 10.6 born and 9.2 reared for the last five years. Deaths, as measured by the difference between numbers born and numbers reared, were 1.4 per litter in both cases. For most of the 1970s, numbers born remained fairly constant at just over ten per litter but weaners reared continued to improve for much of the time as the mortality fell again after rising temporarily to 2.0 deaths per litter.

The improvement in numbers born and reared (1.6 pigs a litter) during this 40 year period appears to have been in part due to better breeding stock and design of farrowing houses but must also have come from better stockmanship. This involves attending to sows at all times and ensuring that they are properly fed during pregnancy. The achievement of larger litters helped to justify investment in better housing, especially for farrowing, to improve conditions and performance. The increase in average size of herd, as recorded in the scheme, from 24 to over 160 sows, shown in Figure 6.2, allowed new developments in style of accommodation and provided cost benefits from economies of scale in size of housing. It also led to the employment of specialists to work full-time on pigs without the distraction of having to regularly undertake other tasks on the farm.

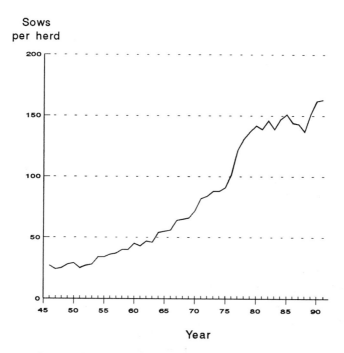

Figure 6.2 Average number of sows per herd

With expansion to larger herds and improved performance, pig production increasingly was being taken more seriously by farmers. From being subsidiary activities absorbing capital and labour left over from primary enterprises, many units became major enterprises in their own right. Herd expansion by those remaining in pigs continued throughout the life of the scheme.

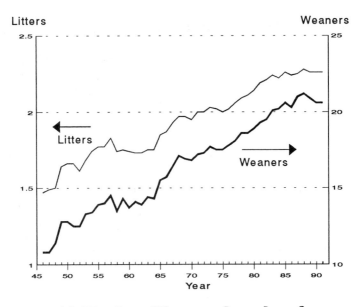

Figure 6.3 Number of litters and number of weaners per sow in herd a year

As litter sizes increased so also did the farrowing frequency from about 1.5 to over 2.2 litters per sow in herd a year. This increase was mainly due to earlier weaning. Until the mid-1960s, the traditional age for weaning young pigs was eight weeks and although with this technique it was possible to achieve two litters per sow a year, very few were this successful. For several years the average was 1.75 and this was the standard most producers aimed for. Initially only a few tried weaning earlier, some at three weeks, but more at five or six weeks of age. A few even tried 10 day weaning, though this system was short-lived due to the many difficulties encountered. By the 1980s most breeders had adopted three, four or five week weaning. With good timing and high conception rates, 2.6 litters per sow a year became possible from three week weaning. Achievements in recent years suggest that the overall standard of just under 2.3 litters a year, which is effected by a few herds weaning later than three weeks, will now be difficult to better. Both the new 2.3 from a possible 2.6 and the old 1.75 from a possible 2.0 give similar success rates of 88 per cent.

From the late 1940s, the average number of weaners produced per sow in herd a year increased almost continuously to peak at 21.2 in 1988. After that, until the end of the scheme in 1991, the number declined marginally to 20.6 a year, due to slightly fewer farrowings and higher mortality. The increase in the number of weaners per sow means that fewer sows are required now to produce a given number of pigs.

Table 6.1 Number of sows required to produce 1,000 weaners a year

Three year average	Litters per sow a year	Weaners per litter	Weaners per sow a year	Sows required to produce 1,000 weaners a year
1949-51	1.65	7.7	12.7	79
1959-61	1.74	8.1	14.0	71
1969-71	1.97	8.6	16.9	59
1979-81	2.15	8.8	18.9	53
1989-91	2.26	9.2	20.7	48

On average for the three-year period of 1989-91 only 48 sows were needed to produce 1,000 weaners a year whereas 79 sows were needed in 1949-51, a reduction over time of nearly 40 per cent. Between these years the number of litters per sow in herd increased from 1.65 to 2.26 a year and the number of weaners from 7.7 to 9.2 per litter. Combined, they raised output by eight weaners a year from 12.7 to 20.7 per sow, an improvement of 63 per cent.

Feed requirements

Another factor which has considerable effect on the costs of producing weaners is the quantity and cost of feed used. To measure feed properly, all feed given to breeding stock must be taken into account; in addition to feed for sows and in-pig gilts, includes that given to boars and any maiden gilts awaiting service, together with creep feed for the young piglets until they reach a standard age, in this case eight weeks. The total feed used by the breeding herd has been divided by the number of weaners produced to give the quantity per weaner as shown in Figure 6.4.

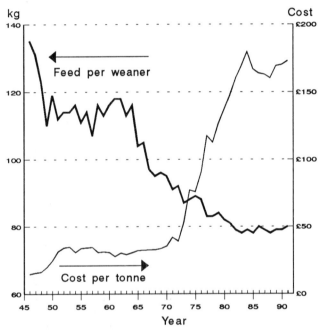

Figure 6.4 Feed used per weaner produced and average cost per tonne

At first glance Figure 6.4 suggests that breeding herd feed was wasted in the earlier years. This was probably true in part, but feedingstuffs were rationed until the early 1950s and balanced feeds were often difficult to obtain. Little specialist feed for the piglets was available; they relied on sow's milk and any sow meal they could sneak from their mothers. Weaners seldom weighed more than 30 lb (14 kg) at eight weeks. There were fewer weaners per sow to share this feed in those days so the amount per weaner was excessive by current standards.

With the introduction of early weaning in the 1960s, piglet feeds, although expensive, became more popular. From then on the total quantity of breeding stock feed increased to around 1.6 tonnes per sow in herd a year, compared to about 1.45 tonnes per sow some ten years earlier. By then, however, the number of weaners produced per sow a year was also increasing, which offset the higher feed requirement to give a better performance in feed used per weaner.

As shown in Table 6.2, since 1980, feed used for piglets has been recorded separately from the sow feed. Approximately 85 per cent of piglet feed was in the form of purchased compounds (starter and creep feeds) and 15 per cent own mixed. For sows purchased compounds have usually been between 60 and 65 per cent of the total.

Table 6.2 Sow and piglet feed used

Year	Sow feed per sow	Sow feed per weaner	Piglet feed per weaner	Sow feed per tonne	Piglet feed per tonne	Piglet weight to 8 weeks
	t	kg	kg	£	£	kg
1980	1.20	63.3	18.7	121	192	17.9
1981	1.19	61.6	19.0	130	206	18.0
1982	1.18	60.4	19.0	139	228	18.2
1983	1.18	59.6	18.6	150	233	18.1
1984	1.21	60.5	18.7	159	246	18.1
1985	1.21	59.6	18.5	146	236	18.2
1986	1.25	61.4	18.6	143	237	18.2
1987	1.27	60.4	18.4	141	235	18.3
1988	1.27	59.8	18.0	137	240	18.3
1989	1.28	61.1	17.5	144	258	18.1
1990	1.28	61.9	17.2	144	264	18.1
1991	1.29	62.7	17.3	147	267	18.2

The quantity of sow feed used has increased recently and in 1991 averaged 1.29 tonnes per sow in herd. This was partly due to more outdoor production, where requirements are about 1.4 tonnes per sow, but maybe also because as seen in the 1950s and 1960s, when the price of sow feed falls it is sometimes used less carefully. The price of sow feed rose almost continuously from 1972 (£37 per tonne) to 1984 (£159), so that when the fall came in the following years it brought relief from the pressures of seemingly ever-increasing prices, causing some relaxation in otherwise fairly strict control over feed use.

On the other hand the price per tonne of piglet feed rose above the 1984 peak during the last three years of the scheme. Compounders seem to introduce new and usually more expensive feeds for piglets frequently. Fortunately some of the increased cost has been offset as the quantity required per weaner has declined, with only minor effect on weaner weights.

Costs per weaner

The increase in the number of weaners produced per sow has helped to contain weaner costs in current terms, especially feed costs. Had weaners in the three years 1989-91 been produced at the same level of efficiency as 20 years earlier, then feed costs would have been £2.50 more per weaner. Labour productivity has also improved by advances in buildings and equipment development. Labour costs per weaner produced did not rise in line with wage rates (as measured by the awards of the Agricultural Wages Board). Not until the Community Charge (Poll Tax) came into force in 1990, which many farmers felt obliged to pay for their employees, and sometimes their spouses, did labour costs increase substantially. Other costs also rose at a similar rate to labour mainly through higher charges for veterinary services, power, water and miscellaneous expenses and fees (Table 6.3).

The average costs of producing weaners changed little between 1950 and 1970. Feed prices were reasonably stable and the small rises in prices of labour and other items were offset by improved production efficiency.

Table 6.3 Average breeding costs per weaner (8 weeks) in real terms[a]

Year[b]	Feed	Labour	Other	Stock depreciation	Total	Annual[c] total
	£	£	£	£	£	£
1970	24.54	6.97	6.90	1.24	39.65	5.46
1971	25.50	6.67	6.47	1.74	40.38	6.05
1972	22.09	6.82	6.27	1.12	36.30	5.85
1973	26.43	7.04	6.07	-.06	39.48	6.90
1974	33.74	7.51	6.16	1.25	48.66	9.72
1975	27.54	7.77	6.87	.74	42.92	10.49
1976	27.15	7.62	7.65	.76	43.18	12.58
1977	28.76	6.58	7.32	1.00	43.66	14.80
1978	25.16	6.64	7.42	.97	40.19	14.90
1979	25.73	6.80	7.63	1.48	41.64	17.14
1980	23.07	6.62	7.21	.72	37.62	18.36
1981	21.63	6.35	7.16	.67	35.81	19.70
1982	20.83	6.21	6.84	.38	34.26	20.74
1983	20.64	6.28	6.63	.71	34.26	21.75
1984	21.33	6.07	6.39	.50	34.29	22.86
1985	18.44	5.98	6.26	.62	31.30	22.10
1986	17.88	6.05	6.21	.84	30.98	22.74
1987	16.82	5.94	5.90	.98	29.64	22.62
1988	15.67	5.73	5.68	1.47	28.55	22.72
1989	15.53	5.69	5.68	.69	27.59	23.62
1990	14.43	5.77	5.94	.68	26.82	25.00
1991	13.82	5.81	5.89	1.12	26.64	26.64

(a) Reflated by the retails prices index to 1991 values
(b) Year ended 30 September
(c) Annual total in current terms

In current terms the total costs per weaner, by eight weeks of age, increased nearly fivefold, from £5.46 in 1970 to £26.64 in 1991, despite almost continual improvement in performance. In real terms, however, breeding costs per weaner have been falling since reaching a peak in 1974. When costs in earlier years are reflated by changes in the Retail Prices Index, to put them in terms of money of 1991 buying power, the total per weaner falls almost continuously. Between 1974 and 1991 total costs per weaner decreased from the record level of £48.66 to £26.64, a fall of 45 per cent. The fall in feed costs from £33.74 to £13.82 was even greater at 59 per cent and, as feed was the major item of costs, was largely responsible for the decline in total costs per weaner. Labour costs peaked in real terms a year later in 1975 and other costs the year after that. By 1991, labour costs per weaner averaged £5.81 and other costs £5.89. Although both had been increasing slightly, they showed falls of 25 and 23 per cent respectively from their highest points.

Stock depreciation is basically the difference between the purchase price and cull value at sale of boars, sows and gilts. A contribution also comes from the declining value of new purchases on hand at the beginning of the recording year that were still in the herd at the end of the year. Short-term fluctuations in market prices of cull breeding stock were ignored for the opening and closing valuations; therefore the majority of sows were valued at the same standard rate on both occasions. Usually the overall average values of breeding stock in established herds, where replacement and culling rates are fairly constant, between opening and closing valuations were little different. It is only in cases where herd sizes have either increased during the year and have a high proportion of new purchased stock on hand, or decreased by sale of some existing stock without replacement, that valuations were likely to differ substantially. Home-bred gilts were transferred to the breeding herd at slightly over bacon pig value and, for valuation purposes, were increased by stages to the standard rate when they became sows. As the transfer value was usually less than the cost of purchased gilts, stock depreciation was lower for herds using home-bred replacements. Comparisons of stock depreciation rates for herds using home-bred gilts, together with the standard of performance achieved by the two groups, are given in Chapter 10. No credit was allowed for deaths so a high sow mortality rate increased depreciation. Total depreciation was then divided by the number of weaners produced to give the charge per weaner.

An indication of the current level of depreciation charges can be ascertained by reference to prevailing costs of new or replacement breeding stock and the value of culls sold. Changes in average prices over the years and the contribution the cull value makes towards the cost of replacement are shown in Table 6.4.

Table 6.4 Average prices of breeding stock[a]

Year	Boars			Sows and gilts		
	Purchases[b]	Sales	Sales as % of purchases	Purchases[b]	Sales	Sales as % of purchases
	£	£	%	£	£	%
1970	71.39	28.75	40	31.62	25.63	81
1971	73.92	21.07	29	31.29	20.29	65
1972	75.56	29.71	39	32.85	23.53	72
1973	78.30	42.63	54	35.57	33.52	94
1974	97.13	40.06	41	46.27	36.34	79
1975	111.78	43.20	39	56.15	41.60	74
1976	136.50	63.42	46	66.26	57.65	87
1977	150.22	65.25	43	68.14	56.05	82
1978	184.77	75.11	41	79.20	68.42	86
1979	208.19	72.03	35	88.27	61.80	70
1980	263.06	82.02	31	95.25	78.42	82
1981	271.70	89.99	33	97.74	79.95	82
1982	306.70	100.16	33	108.72	95.99	88
1983	327.04	80.42	25	106.24	87.74	83
1984	326.69	80.26	25	112.49	92.26	82
1985	365.88	83.53	23	125.78	97.49	78
1986	385.42	79.79	21	130.27	93.18	72
1987	412.13	86.10	21	131.97	91.58	69
1988	439.79	77.71	18	139.86	86.65	62
1989	467.00	102.97	22	137.12	105.59	77
1990	507.12	124.36	25	147.43	113.80	77
1991	532.00	107.72	20	149.73	102.84	69

(a) Includes any young stock purchased or sold
(b) Excludes home-bred stock

The cost of purchased boars has risen more steeply than purchased gilts. In 1991 the boars averaged £532 each, some seven and a half times the cost in 1970, whereas gilts were nearly £150 each in 1991, four and three-quarters times the 1970 cost. With both the boars and the gilts there were wide variations around these average figures for individual herds.

Prices received for cull boars and sows sold also varied greatly and fluctuated from year to year according to supply and demand. Over this period many cull sows were exported to Germany for manufacturing pigmeat products and prices were often influenced by currency exchange rates and supplies from elsewhere. The British prices also reflected the profitability of home production and prevailing prices paid for finished pigs. 1988 was a bad year when low prices for cull breeding stock were further depressed by additional slaughterings from those giving up pig production.

The price received for cull breeding stock sold as a percentage of the cost of replacements has been declining, especially for boars. For the five years 1970-74, the average price received for boars sold was 41 per cent of the cost of a new one, whereas for 1987-91 the cull value only provided 21 per cent of the replacement cost. For sows and gilts, where nearly all purchases were gilts and sales were sows, the value of culls sold contributes far more towards the cost of replacements but here too the proportion has been declining. In 1970-74, the average price received formed 78 per cent of the replacement cost, compared with 71 per cent for 1987-91.

In addition to the purchased gilts listed in Table 6.4 a further intake comes from home-bred gilts. About a third of the total brought into the breeding herd have been home-bred, though more recently this share has been falling. Practically all boars were purchased with only an occasional one or two home-bred by the larger herds. For sales there have usually been some 15 times as many cull sows sold by herds in the scheme as cull boars.

Results for 1991

The average results of the breeding herds in the scheme in 1991 are shown in more detail in Table 6.5. To illustrate the wide variation between herds in nearly all the production factors, results for the best 20 and worst 20 herds are also given.

On average, 2.26 litters were farrowed per sow in herd from weaning at 26 days, with 10.6 born and 9.1 reared per litter, to produce 20.6 weaners (weighing 18.1 kg at eight weeks) per sow a year. Sow feed used averaged 1.29 tonnes per sow, or 62.7 kg per weaner, costing £147.14 per tonne. The weaners consumed, by eight weeks of age, 17.3 kg of piglet feed each at £266.86 per tonne. The sow and piglet feed combined cost £13.82 per weaner produced. Labour and other costs, detailed in Table 6.5, brought the total costs to £26.64 per weaner.

Table 6.5 Average and range of breeding results 1991

Breeding stock	Average	Best 20[a]	Worst 20[a]
Number of herds	119	20	20
Number of sows in herd[b]	163	159	117
Number of litters	370	375	258
Litters per sow in herd	2.26	2.35	2.21
Age at weaning (days)	26	24	28
Live pigs born per litter	10.6	10.9	10.3
Weaners per litter	9.1	9.5	8.5
Weaners per sow in herd	20.6	22.4	18.7
Weight of weaners at 8 weeks	18.1 kg	18.0 kg	18.1 kg
Per cent sold as weaners/stores	39 %	51 %	49 %
Culled sows percentage	37 %	34 %	44 %
Gilts brought in - % of herd	42 %	42 %	43 %
Gilts brought in - % purchased	73 %	63 %	74 %
Sow feed used per sow in herd	1.29 t	1.25 t	1.35 t
Sow feed per weaner to 8 wks	62.7 kg	55.6 kg	72.3 kg
Piglet feed per weaner to 8 wks	17.3 kg	17.2 kg	17.9 kg
Cost of sow meal per tonne	£147.14	£146.45	£147.73
Cost of piglet meal per tonne[c]	£266.86	£258.65	£286.51
Compounds as % of total meal	70 %	64 %	67 %
Costs per weaner at 8 weeks	£	£	£
Feed	13.82	12.60	15.79
Labour	5.81	4.99	7.28
Other costs			
Farm transport	.43	.33	.77
Vet. and Vet. supplies	.82	.74	1.21
A.I. fees	.11	.05	.20
Power and water	1.13	.90	1.43
Miscellaneous expenses	.63	.55	.78
Litter	.32	.23	.43
Maintenance	.58	.50	.58
Equipment charge	.29	.23	.42
Buildings charge	1.50	1.24	1.68
Pasture charge	.08	.02	.13
Total other costs	[5.89]	[4.79]	[7.63]
Stock depreciation	1.12	0.63	1.70
Total breeding costs	26.64	23.01	32.40

(a) Selected on total costs per weaner for breeding (adjusted for weight variation) feeding
(b) Monthly average (including in-pig gilts)
(c) To eight weeks of age

Performance ranged widely. The best 20 herds produced more litters (2.35) through an average weaning age of 24 days. These were larger litters, both at birth (10.9) and at eight weeks (9.5). They averaged 22.4 weaners per sow, weighing virtually the same (18.0 kg) as the overall average. Sows in the best herds used less feed (1.25 tonnes a year) and further benefited from producing more weaners to share this overhead, so that only 55.6 kg of sow feed (at £146.45 per tonne) was used per weaner. Piglet feed per head used by these herds at 17.2 kg was virtually the same as the average of all herds but at £258.65 per tonne it was cheaper, in part because slightly more was home-mixed. In total, feed cost £12.60 per weaner, £1.22 less than average. Labour and other costs were also lower and altogether costs amounted to £23.01 per weaner (£3.63 less).

The worst 20 herds produced fewer litters (2.21) with a 28 day average weaning age. The litters were smaller, 10.3 born and 8.5 reared, to produce only 18.7 weaners per sow a year. More sow feed was used, 1.35 tonnes per sow, or 72.3 kg per weaner, costing £147.73 per tonne. These weaners used more piglet feed (17.9 kg) at a higher cost of £286.51 per tonne. Feed costs were therefore £15.79 per weaner, £1.97 above average. Higher labour and other costs, brought the total to £32.40 per weaner (£5.76 more than average).

Between the best 20 and the worst 20 herds the difference in total costs came to £9.39 per weaner. This was largely in respect of the number of weaners per sow, feed use and different labour and other costs. The greater number of weaners produced by the best herds provided a better spread of costs to give a much lower figure per weaner. The cost of feed per weaner for the worst herds was 25 per cent more than for the best herds. Labour costs were higher by 46 per cent and other costs 59 per cent. Stock depreciation was 170 per cent higher due to greater sow mortality (5.5 per cent compared to 3.5), lower prices received for cull breeding stock and higher prices paid for replacement gilts.

Variation in performance

The costs of a breeding herd must be borne by the weaners produced. As most of these costs are incurred irrespective of the number of weaners reared, it follows that herds rearing large numbers of weaners per sow in herd a year have more pigs over which to spread costs and so achieve a lower cost and, other things being equal, a higher profit margin per weaner.

The quantity and cost of feed, as well as labour and other costs, vary considerably between individual herds. The dispersion around the average of some of these results is worth examination to show the wide variation that exists.

Figure 6.5 shows the relationship between the number of litters (farrowing frequency) and the number of weaners produced per sow in herd in 1991. Each dot represents one herd. For individual herds, litters per sow ranged from 1.5 to 2.5 per year and the number of weaners from 11 to 26.6 per sow. Most herds were grouped around the average for each factor. Some 87 per cent of herds achieved between more than 2.1 and less than 2.5 litters per sow, while 75 per cent produced more than 18 and less than 23 weaners per sow. The overall averages were 2.26 litters and 20.6 weaners per sow.

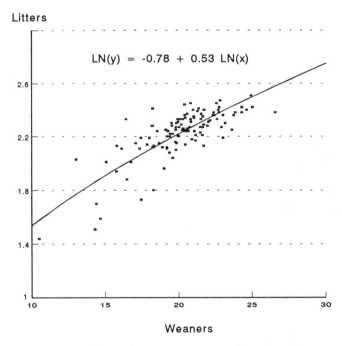

Figure 6.5 Litters per sow related to the
number of weaners produced per sow 1991

The trend line shows that as the number of litters per sow a year increases, so also does the number of weaners per sow. A high output of weaners is, therefore, more likely to be achieved in those herds producing more than 2.1 litters per sow a year.

The quantity of breeding stock feed used depends on the number of weaners produced per sow. The total sow and piglet feed per weaner falls considerably as the number of weaners per sow in herd a year increases, as in Figure 6.6.

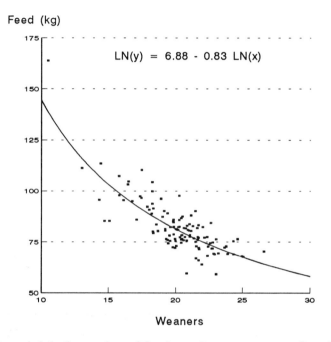

Figure 6.6 Quantity of feed used per weaner related to the number of weaners produced per sow 1991

Here, the range extends from 60 to over 110 kg of feed per weaner, with 70 per cent of the herds falling between 70 to 90 kg. The average was 80 kg.

Similarly, the total costs per weaner at eight weeks of age decline as weaner numbers per sow increase, as shown in Figure 6.7. Nearly all herds came within a range of £20 to £35 per weaner and 60 per cent of them were between £24 and £29 each. The total costs for all herds averaged £26.64 per weaner.

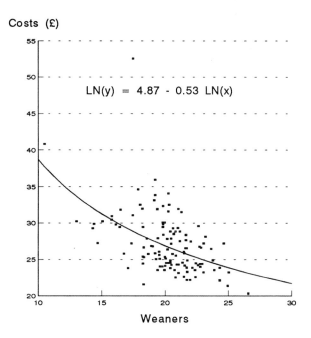

Figure 6.7 Total costs per weaner related to the number of weaners produced per sow 1991

These scatter graphs show the importance of weaner numbers per sow a year in affecting the quantity of feed and the total costs per weaner. The higher numbers provide a better distribution of feed and costs to improve efficiency and reduce costs per weaner.

The relationship between the capital value of buildings and equipment per sow and the number of weaners is shown in Figure 6.8. The former covers all buildings and equipment (including any milling and mixing plant) and also tractors, other vehicles, balers and bale handling equipment, when used exclusively for pigs but not for stock or working capital for feed, labour and other costs. Herds with high capital investment were likely to produce more weaners per sow than those with lower investment.

69

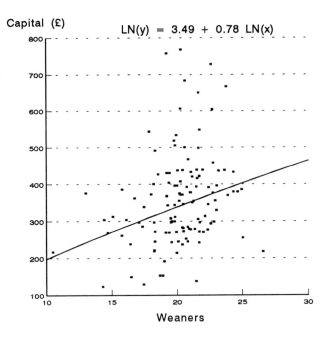

Figure 6.8 **Capital value of buildings and equipment per sow related to the number of weaners produced per sow 1991**

The capital value of buildings and equipment per sow in 1991 varied greatly, from £123 to £768 per sow, around an average of £370 for all herds. For just over 72 per cent of herds it lay between £250 and £500 per sow.

The scatter graphs show that some 15 per cent of breeding herds produced less than 18 weaners per sow in 1991. Some low numbers have been caused by health problems, or restocking of the herd, but in several herds improvements would be possible simply by ensuring that sows were served promptly to produce more litters a year. Some increase in litter size and survival rate would also be valuable.

The possibilities of achieving different numbers of weaners per sow by changing farrowing intervals and weaners per litter are shown in Table 6.6.

Table 6.6 Weaners per sow a year from varying litter size
and number of litters per sow

Litters per sow a year	Farrowing interval	Weaners per litter		
		8	9	10
	Days	Number of weaners		
1.8	203	14.4	16.2	18.0
2.0	183	16.0	18.0	20.0
2.2	166	17.6	19.8	22.0
2.4	152	19.2	21.6	24.0
2.6	140	20.8	23.4	26.0

Two litters of nine pigs per litter gives 18 weaners per sow a year. An extra pig per litter gives 20 weaners but increasing litter size is not easy. Nearly the same improvement could be achieved, with no increase in litter size, just by producing 2.2 litters per sow instead of 2.0 a year. In some herds, this improvement is possible without changing the age at weaning by making sure that sows are served properly at the right time. This avoids unproductive sows having to be carried by the rest of the herd and so depressing performance.

As the number of litters per sow a year increases, the interval between farrowing becomes progressively shorter. Between 1.8 and 2.0 litters per sow a year, the farrowing interval is reduced by 20 days but between 2.2 and 2.4 it is down to 14 days. Once a reasonable standard has been achieved it becomes increasingly more difficult to improve the number of litters per sow, as there is less time available.

The savings in cost of producing more weaners per sow a year are considerable, as shown in Table 6.7. In this case, costs cover the young pigs up to a sale weight of 30 kg.

The range of 18 to 22 pigs produced per sow a year cited in Table 6.7 covered 75 per cent of the breeding herds in the scheme in 1991. Most of the costs per sow over this range do not vary much with the number of pigs produced. The only additional charges to be incurred as weaner numbers increase would be for extra feed for the sow during lactation, piglet feed and veterinary costs (included with other costs) mainly for routine medication for each weaner.

Table 6.7 Effect of the number of pigs produced per sow a year on the profitability of weaners (30 kg)

Number of pigs per sow a year	18	20	22
Costs per sow	£	£	£
Feed per sow[a]	146	146	146
Extra sow feed during lactation[b]	18	20	22
Share of boars' feed[c]	8	8	8
Share of replacement gilts' feed[d]	8	8	8
Feed for young pigs [e]	166	184	202
Total feed	346	366	386
Labour	135	135	135
Other costs	135	136	137
Stock depreciation	20	20	20
Total costs	636	657	678
Value of weaners (@ say £36 each)	648	720	792
Margin per sow	12	63	114
Average costs per weaner	£35.33	£32.85	£30.83
Margin per weaner	£0.67	£3.15	£5.18

(a) 2.5 kg per day @ £160 per tonne.
(b) 0.25 kg per pig a day for 25 days @ £160 per tonne.
(c) 1 tonne per sow @ £160 per tonne shared by 20 sows.
(d) One-third of sows replaced each year. Say 60 days awaiting service at 2.5 kg per day @ £160 per tonne.
(e) 40 kg each @ £230 per tonne.

Total costs are assessed at £636 a year for a sow producing 18 pigs, £657 for 20 pigs and £678 for 22 pigs. On a per pig basis these costs averaged £35.33, £32.85 and £30.82 respectively. If they were valued at £36 a head, then 18 pigs per sow a year leaves a small margin of 67p per weaner, 20 per sow one of £3.15 while 22 per sow achieves a quite respectable £5.18 per weaner.

Costs per weaner are also influenced by the quantity of feed used, especially sow feed. The quantity of sow feed used by herds in the scheme in 1991 ranged from 1.0 to 1.5 tonnes per sow in herd. Although this includes a share of the boar's feed and any for gilts awaiting first service, most of the difference comes from variation in the amount used by the established sow herd. It does not include any piglet feed for the young pigs. The extent of this range clearly suggests that extravagance or wastage does occur in some herds, most likely during the dry sow stage. The effect of varying the quantity of sow feed used (excluding piglet feed) over a shorter range of 1.1 to 1.4 tonnes per sow on costs and margins is shown in Table 6.8.

Table 6.8 Effect of the quantity of feed per sow a year on the profitability of weaner production

(20 weaners averaging 30 kg per sow a year)

Quantity of feed per sow (kg)	1,100	1,200	1,300	1,400
Costs per sow a year	£	£	£	£
Sow's feed @ £160 per tonne	176	192	208	224
Remainder of costs*	475	475	475	475
Total costs	651	667	683	699
Value of weaners @ £36 each	720	720	720	720
Margin per sow a year	69	53	37	21
Average cost per weaner	£32.55	£33.75	£34.15	£34.95
Margin per weaner	£3.45	£2.65	£1.85	£1.05

* Includes feed for young pigs (£184), stock depreciation (£20), labour (£135) and other costs (£136)

Only the quantity of feed used per sow varies. The remaining costs are held constant in this example, though it is possible that labour may rise as more feed is handled. As more feed is used both costs per sow and per weaner rise and margins fall. At the level of costs used here, £160 per tonne, each 100 kg of feed used a year adds £16 to costs per sow and 80p to costs per weaner. If feed were more expensive the increase to costs would be correspondingly higher. It is important, therefore, that strict control is exercised over this expensive commodity as the savings this could produce are well worthwhile.

In the breeding stage of production, the number of weaners reared per sow and the quantity of sow feed used account for much of the variation in profitability between herds. In the examples in Tables 6.7 and 6.8, the remaining costs were kept constant but in reality these may also vary, usually by small amounts and often in association with other resources available to the unit. The quantity of piglet feed per weaner often varies, sometimes reflecting growth rates. Likewise the cost of feed per tonne, where home-mixed is likely to be cheaper than purchased compounds. Labour requirements may vary due to work routine and the design and layout of buildings. Rates of pay vary considerably, as do the skill and ability of the persons involved. It is not uncommon to find a pig unit where labour achieves good results with apparent ease, while a similar work-force and size of unit elsewhere seem to have difficulty in coping. With so much variation between herds, the emphasis must be on recording to establish the levels being achieved and how they compare with other herds.

Chapter 7

Feeding Herd Performance

The feeding, or finishing, stage of pig production in the scheme covered the period to sale from the time the young pigs were eight weeks of age, if home-bred, or from the time they were brought in, if purchased. Any gilts retained for breeding stock replacements were included in the feeding herd until transfer back to the breeding herd, usually at about 90 kg liveweight.

Herd size

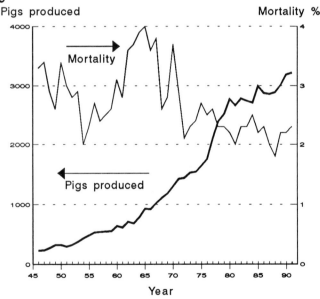

Figure 7.1 Average number of pigs produced per herd a year and mortality percentage

The average number of pigs produced per herd increased almost every year from 1946 to 1991 (Figure 7.1) in line with the national trend.

Compared with the five years of 1946-50, when the number produced averaged 276, production for 1987-91 increased elevenfold to 3,031 per annum. Mortality of feeding stock was variable over the years but was generally at a lower level after 1972. In 1946-50 the mortality rate averaged 3.1 per cent compared with 2.1 per cent for 1987-91.

Liveweights in and out

The average liveweight of pigs produced fell during the second half of the 1950s, when controls were relaxed. Then pigmeat supplies became more plentiful and more pigs were required for slaughter at lighter weights. The increased popularity of heavy pigs in the 1960s raised average weights but they fell back during

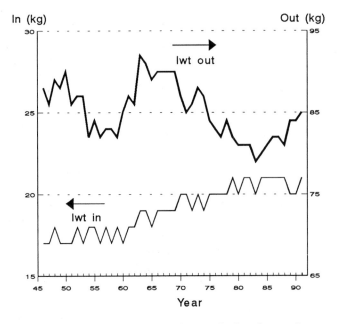

Figure 7.2 Average liveweight of pigs in and out
of the feeding herd

76

the 1970s as the escalation in feed prices made heavy pig production expensive and less profitable. More recently, many buyers have been seeking slightly heavier pigs for the pork and cutter trade, causing weights of finished pigs to rise once again.

Although the liveweight of home-bred young pigs entering the feeding herd at eight weeks did rise marginally, the average weight of pigs brought in increased mainly because traded weaners became heavier. In the early years it was common for most weaners to change hands at weaning (then at eight weeks) but during the 1960s buyers, aiming to reduce stress and mortality, sought stronger pigs often of ten weeks old when they weighed around 25 kg each. Weights of traded weaners further increased during the 1970s and since then many have been about 30 kg. Since 1970 the combined liveweight of young pigs brought in, including both home-bred and purchased weaners, has remained fairly constant at about 20 kg (Figure 7.2).

Feed requirements and costs

The feed conversion rate has been influenced by the changes in weights of pigs entering and leaving the feeding herd but, nevertheless, substantial improvement has been achieved over the years. Feed conversion rates measure the quantity of feed used to increase the liveweight of pigs (kg of feed per kg liveweight gain). Other feeds were converted to a meal equivalent (see Appendix 4) and added to meal. Liveweight gain is the difference between sales (and transfers out) and purchases (and transfers in), adjusted for weights of pigs on hand at the beginning and end of the recording period.

During the recovery period following the wartime rundown of pig production, the feed conversion rates reflected the poor quality of feed available and were indeed pretty grim, averaging 4.83 for the seven years 1946-52. Since then improvement has been steady and consistent throughout the life of the scheme, with only the occasional minor set-back; see Figure 7.3. By the last seven years 1985-91, the average conversion rate was down to 2.77 - just 57 per cent of the immediate post-war standard.

The cost of feed per tonne rose after the war until 1952, then changed little over the next 20 years. Between 1972 and 1984, costs increased sharply from an average of £33 per tonne to

£159, mainly as a result of UK entry to the EC and high inflation. In 1985 feed costs fell back some £12 per tonne and remained close to this level until the end of the scheme.

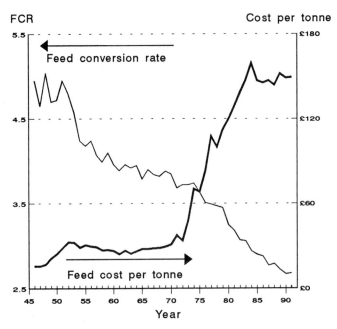

**Figure 7.3 Average feed conversion rate
and cost of feed per tonne**

For 1985-91 the cost of feed averaged £147.64 per tonne and the conversion rate 2.77 to give a feed cost of 40.9p per kg liveweight gain. Had the conversion rate not improved and remained at the 1946-52 level of 4.83, the cost of feed at 1985-91 prices per tonne, would have been 71.3p per kg liveweight gain - 74 per cent more than the actual cost with the 2.77 conversion rate. With the addition of labour and other costs, total costs would have amounted to some 82p per kg liveweight, which is equal to 109p per kg deadweight and 10p more than the average net price received for all finished pigs sold in 1991.

Between 1955 and 1970 the improvement in feed conversion was considerably lower than in other periods. With hindsight this was a good time for pig producers. The guaranteed scheme ensured reasonable pig prices and with feed prices stable, the financial pressures on producers were few and there was not the urgency to improve. Some improvement in conversion was made from 4.18 in 1955 to 3.85 in 1970 but this was quite small compared with later improvements. Changes averaged over three years are shown at ten-year intervals in Table 7.1.

Table 7.1 Improvements in feed conversion rates

Three year average	Conversion rate	Improvement
1949-51	4.79	-
1959-61	3.98	.81
1969-71	3.80	.18
1979-81	3.29	.51
1989-91	2.69	.60

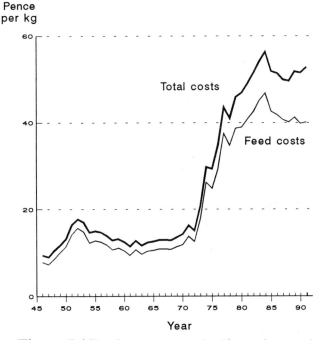

Figure 7.4 Feeder costs per kg liveweight gain

The improvement between 1959-61 and 1969-71 of only 0.18 was far less than that achieved in the two following decades, when conditions for producers were more difficult. The constancy of feed and total costs per kg liveweight gain in the late 50s and 60s is illustrated in Figure 7.4. After this, both increased dramatically as the EC cereal regime and inflation took hold.

After many years of close similarity the feed cost and total cost lines started to diverge in the late 1970s and have continued doing so since. The difference between the two covers labour, other costs and the mortality charge. The mortality charge is the value of weaners brought into the feeding herd which subsequently die. While feed costs per kg liveweight gain have declined from the peak of 1984, these remaining costs continued to rise and partly offset some of the benefit. By 1991, labour, other costs and the mortality charge together amounted to 31 per cent of feed costs compared to only 20 per cent in 1984. Most of this increase came during the last three years from 1989 onwards, as virtually all of these costs incurred in feeding stock went up considerably. Details of average costs per kg liveweight gain from 1970 are given in Table 7.2.

As in the breeding stage, the current costs of feeding stock production, shown in Figure 7.4, changed little between 1950 and 1970. Feed prices remained reasonably stable and the small increases in labour and other costs were offset by the improvement in the efficiency of feed conversion. Since 1970, the current value of annual total feeding costs has risen sharply, despite further improvements in conversion rates, reaching a peak in 1984 of 56.3p per kg liveweight gain, almost a fourfold increase in just 14 years. Total costs then fell for the next four years, as feed costs per tonne declined and conversion rates continued to improve, but rose again during the final three years to 1991 mainly through higher labour and other costs to reach an average of 52.8p per kg for 1991. Between 1970 and 1991, labour, other costs and the mortality charge increased by 429 per cent in current values, while feed costs per kg rose by only 237 per cent.

The structure of costs throughout has been dominated by feed but the proportion of labour and other costs has been increasing, especially in the later years. For most of the 1980s labour and other costs formed 15 to 16 per cent of total costs but by 1991 they increased to 22 per cent. This further underlines the added importance of labour and other costs in present day total costs of pig production.

Table 7.2 Average feeding costs per kg liveweight gain in real terms[a]

Year[b]	Feed	Labour	Other	Mortality charge	Total	Annual[c] total
	p	p	p	p	p	p
1970	86.4	8.0	7.2	2.2	103.8	14.3
1971	92.1	7.4	7.3	2.0	108.8	16.3
1972	78.2	7.4	6.8	1.3	93.7	15.1
1973	102.4	8.0	7.5	1.7	119.6	20.9
1974	131.2	8.5	7.0	2.0	148.7	29.7
1975	101.5	8.6	7.8	2.0	119.9	29.3
1976	99.9	8.2	8.6	2.1	118.8	34.6
1977	110.3	7.1	8.8	2.1	128.3	43.5
1978	93.6	6.8	8.6	1.6	110.6	41.0
1979	94.0	7.3	8.5	1.7	111.5	45.9
1980	79.7	6.8	8.0	1.6	96.1	46.9
1981	74.2	6.5	7.1	1.3	89.1	49.0
1982	70.2	6.5	6.9	1.5	85.1	51.5
1983	71.0	6.2	6.8	1.4	85.4	54.2
1984	70.2	5.9	6.8	1.6	84.5	56.3
1985	60.3	5.7	6.2	1.3	73.5	51.9
1986	57.0	5.4	6.3	1.3	70.0	51.4
1987	53.3	5.2	5.8	1.1	65.4	49.9
1988	50.4	5.1	5.9	0.9	62.3	49.6
1989	48.1	5.1	6.2	1.1	60.5	51.8
1990	42.7	5.2	6.3	1.1	55.3	51.5
1991	40.1	5.3	6.3	1.1	52.8	52.8

(a) Reflated by the retail index to 1991 values
(b) Year ended 30 September
(c) Annual total in current terms

In real terms, total feeding costs per kg liveweight gain peaked in 1974 at 148p in money of 1991 buying power. Apart from a few fluctuations total costs have been falling in real terms since then and by 1991 were only 35 per cent of the 1974 record level. Individual costs in 1991 showed feed at 30 per cent, labour at 61 per cent and other costs at 71 per cent of their respective highest years.

Results for 1991

The average feeding stock results for all herds in the scheme in 1991 are given in Table 7.3 together with the results for the best 20 and worst 20 herds.

Table 7.3 Average and range of feeding stock results 1991

Feeding stock	Average	Best 20[a]	Worst 20[b]
Number of herds	113	20	20
No. of pigs produced per herd	3,217	3,387	2,457
Liveweight of pigs produced	85 kg	86 kg	85 kg
Liveweight of pigs brought in	21 kg	17 kg	28 kg
Pigs brought in - % purchased	32 %	14 %	59 %
Daily liveweight gain	.64 kg	.71 kg	.60 kg
Mortality percentage	2.3 %	2.3 %	2.7 %
Feed conversion rate	2.68	2.39	3.02
Cost of meal per tonne	£153.41	£152.15	£163.10
Cost of feed per tonne^2	£149.38	£148.39	£162.77
Compounds as % of total feed	36 %	10 %	88 %
Costs per kg liveweight gain	p	p	p
Feed	40.1	35.5	49.2
Labour	5.3	4.1	6.1
Other costs			
Farm transport	.6	.6	.7
Vet and vet supplies	.3	.2	.5
Power and water	.7	.5	.7
Miscellaneous expenses	.9	.9	1.1
Litter	.6	.6	.6
Maintenance	.7	.7	.8
Equipment charge	.5	.3	.6
Buildings charge	2.0	1.5	2.3
Total other costs	[6.3]	[5.3]	[7.3]
Mortality charge	1.1	0.9	1.6
Total feeding costs	52.8	45.8	64.2

(a) Selected on total costs per kg liveweight gain
(b) Includes other feeds (mainly by-products) 'converted' to meal equivalent

In terms of total costs per kg liveweight gain the perform-ance of the best 20 herds was 13 per cent better than average, while that of the worst 20 was 22 per cent below. The best achieved a far superior feed conversion rate and daily liveweight gain. Feed costs per tonne were similar to the average of all herds but as less feed was required, the cost per kg liveweight was only 35.5p. Labour and other costs were also lower than average to keep total feeding costs down to 45.8p per kg, a saving of 7p per kg liveweight.

The worst 20 herds only managed a conversion rate of 3.02 despite using more expensive feed, 88 per cent of which was compounds, which gave a feed cost per kg liveweight gain of 49.2p. A much larger proportion of pigs brought in (59 per cent) were purchased at higher weights and this would have had some influence on the poor conversion rate. This group had a lower daily liveweight gain and a higher mortality rate. Labour and most other costs were also higher and total feeding costs amounted to 64.2p per kg liveweight gain, 11.4p more than average and 18.4p more than the best 20 herds.

Variation between herds

As the liveweight of pigs increases, the feed conversion rate is expected to worsen. The scatter of conversion rates related to liveweight, Figure 7.5, around the average 2.68 at 85 kg was considerable. Five herds producing pigs under 60 kg, much light-er than others, were omitted from the total of 113 herds as their inclusion would have distorted the expected trend line. The remainder with average liveweights of pigs produced between 66 and 100 kg clearly demonstrate the wide variation that exists in feed conversion rates between herds.

Feed conversion rates are expected to improve with the quality of feed. It was not possible to measure the quality of feed but the cost per tonne would go some way to reflect this though it may not always apply when the comparison includes both home-mixed and purchased compound feeds.

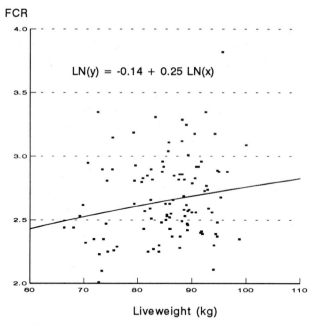

FCR

LN(y) = -0.14 + 0.25 LN(x)

Liveweight (kg)

Figure 7.5 Feed conversion rate related to liveweight of pig produced 1991

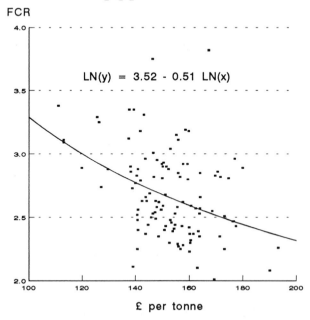

FCR

LN(y) = 3.52 - 0.51 LN(x)

£ per tonne

Figure 7.6 Feed conversion rate related to cost per tonne 1991

Herds with the lowest feed costs per tonne in Figure 7.6 used some inexpensive by-products, while most of the highest costs were found in herds using purchased compounds entirely, where price per tonne was often influenced by the size of the order. Feed costs per tonne ranged from under £120 with conversion rates of over 3.0, to £190 with conversions of less than 2.3. The trend line in the graph clearly indicates that the conversion rate improves as feed becomes more expensive.

As mentioned elsewhere however, it is the cost of feed per kg liveweight gain that is all-important in determining feeding efficiency. While some producers are content to have a relatively poor conversion rate provided the feed is cheap, others justify using expensive feed because they achieve a good conversion rate. Figure 7.7 relates the feed conversion rate to total costs of production.

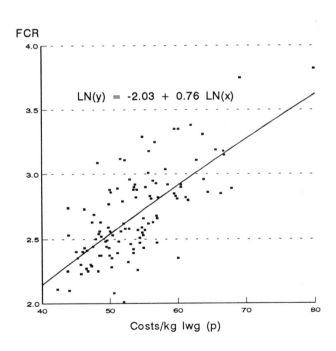

Figure 7.7 Feed conversion rate related to total costs
per kg liveweight gain 1991

Total costs per kg liveweight gain were scattered widely and ranged from a low of 42p, with a 2.11 conversion rate, to a high of 80p, with a 3.82 conversion. Here again, the trend line clearly indicates that total costs per kg liveweight gain rise as the feed conversion rate worsens.

Some relationship might have been expected between total feeding costs per kg liveweight gain and the capital value of buildings and equipment in use. Here again, as with the breeding herds, the two seemed to have little relationship and this may be partly due to the way old buildings were valued. A few herds may have been affected from operating at under-capacity but, with or without them, the graph (Figure 7.8) shows no relationship.

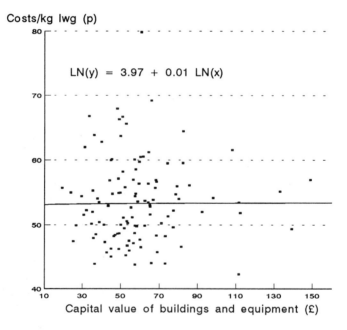

**Figure 7.8 Total costs per kg liveweight gain related
to capital value of buildings and equipment per feeder 1991**

The capital value of feeding stock buildings and equipment ranged from £20 to £149 per feeder. For both of these herds, total costs per kg liveweight gain were fairly similar at 56p

and 57p respectively. When herds expand, or acquire new buildings, the older less efficient buildings are frequently retained, so
that many of today's pig units use a mixture of old and new
accommodation and this is reflected in the overall herd performance. It is perhaps not surprising, therefore, that a survey which
includes all pigs on the farm produces a wide scatter of results
relating costs to the value of buildings and equipment. For most
it is the skillful use of buildings rather than their value which
influences their contribution to production efficiency. Greater
influence on production costs is likely to come from labour,
where the skills and ability of those involved count for more than
the actual buildings.

Production efficiency

The most important factors affecting costs of production and
profitability in the feeding herd are:

1.   The feed conversion rate
2.   The cost of feed per tonne

The feed conversion rate measures the quantity of feed used per
kg liveweight gain for the pigs. Feed conversion rates are con-
ventionally calculated by liveweight to cope with the weight of
pigs in the herd at the beginning and end of each recording
period and young pigs brought live into the herd during the
period. Any pigs sold liveweight through auction markets are
included at their sale weight and others sold deadweight after
slaughter are converted to liveweight according to the formula
(given in Appendix 3). Combining the conversion rate and cost
per tonne gives the cost of feed per kg liveweight gain and, in
many cases, provides a better indication of performance for
comparative purposes. Long-term improvements in costs,
however, become imprecise because of inflation, and feed conversion rate, a physical measurement, is of greater value for this
purpose.

Feed

Several factors contribute to influence the cost of feed per tonne.

They include:

1.   The quality of the feed
2.   The quantity bought at one time
3.   Whether pellets or meal
4.   Bought in bulk or bags
5.   The inclusion of growth promoters
6.   Discounts for prompt payments
7.   Whether purchased compounds or home-mixed
8.   Use of by-products

Some pig producers use high quality feed, which is usually more expensive per tonne, and aim for good performance with low conversion rates. Others go for cheap feed, often using whatever by-products are available, such as skimmed milk, wheat starch, etc, and are quite prepared to accept a poorer conversion rate, providing the cost per tonne is low.

The feed conversion rate is best for small pigs (say 2.0 kg of feed per kg liveweight gain) and steadily worsens as they grow (perhaps 3.0 by 70 kg liveweight). Often in the later stages, the poorer conversion rates are partly offset by the use of cheaper feed with a lower protein content.

Feed costs per kg liveweight gain

Feed is by far the largest item of costs of finishing pigs and, although as labour and other costs increased, it fell from 85 per cent of total production costs in the 1980s, it still formed on average 76 per cent in 1991. Similar feed costs per kg liveweight gain can be achieved from various combinations of price per tonne and conversion rates. Table 7.4 demonstrates the importance of both in determining the cost of feed per kg liveweight gain.

As the conversion rate worsens, the cost of feed per kg liveweight gain increases for all prices per tonne. With feed at £150 per tonne (15p per kg), every 0.1 in the conversion rate is worth 1.5p per kg liveweight gain, or 3p for 0.2 in conversion rate as given in Table 7.4. The table shows what happens to the cost of feed per kg liveweight gain, when the price per tonne changes while the conversion rate is maintained. At a 2.6 rate, a £10 per tonne increase adds 2.6p to the cost of feed per kg liveweight gain. The table also illustrates how the same costs of liveweight gain can be achieved from comparatively low cost feed per tonne converting poorly as from more expensive feed converting efficiently. At £130 per tonne, a 3.4 conversion rate gives the same cost of 44.2p per kg liveweight gain as £170 per tonne with a 2.6 conversion rate. Another example would suggest that a conversion rate of 2.8 is far superior to one of 3.2 but if the feed price is £160 per tonne compared with £140, then both have the same cost of 44.8p per kg liveweight gain.

Table 7.4 Cost of feed per kg liveweight gain at different conversion rates and price per tonne

Conversion rate	Feed price per tonne				
	£130	£140	£150	£160	£170
	p	p	p	p	p
2.2	28.6	30.8	33.0	35.2	37.4
2.4	31.2	33.6	36.0	38.4	40.8
2.6	33.8	36.4	39.0	41.6	44.2
2.8	36.4	39.4	42.0	44.8	47.6
3.0	39.0	42.0	45.0	48.0	51.0
3.2	41.6	44.8	48.0	51.2	54.4
3.4	44.2	47.6	51.0	54.4	57.8

Savings from improvements in the feed conversion rate can amount to very worthwhile sums of money, especially when measured on a per pig basis. This is demonstrated in Table 7.5, which is based on feeding 18 kg weaners to 90 kg finishers, a liveweight gain of 72 kg.

At £150 per tonne, an improvement of 0.1 in the conversion rate of feeding 18 kg weaners to 90 kg finishers is worth £1.08 per pig. A greater improvement of 0.3 is worth £3.24. For a herd producing 3,000 pigs a year, the latter amounts to savings of £9,720 a year and might soon pay for any building alterations needed to obtain this level of performance.

Improvements in feed conversion rates overall are becoming progressively harder to achieve as the standard reaches higher levels. The wide variation that still exists between individual herds shows that for some there remains an opportunity to do better. Pig producers have little control over prices of feed but there are marked differences in the cost per tonne, depending on quality and on whether purchased compounds or feed milled and mixed on the farm is used. Some of the larger herds use by-products, such as wheat starch and milk products, although the latter is not as readily available nowadays as previously. The herds in the Cambridge scheme have for several years provided an opportunity to compare performance results for the different systems. This is dealt with in Chapter 11.

Table 7.5 Value per pig at different levels of improvement in
feed conversion rate and price per tonne
(from 18 kg weaners to 90 kg finishers)

Improvement in feed conversion rate	Feed price per tonne				
	£130	£140	£150	£160	£170
	£	£	£	£	£
0.1	.94	1.01	1.08	1.15	1.22
0.2	1.87	2.02	2.16	2.30	2.45
0.3	2.81	3.02	3.24	3.46	3.67
0.4	3.74	4.03	4.32	4.61	4.90
0.5	4.68	5.04	5.40	5.76	6.12

Chapter 8

Weaner Production and Marketing

Selling weaners

It is estimated that about 30 per cent of the breeding herds in the country sell the progeny produced as weaners or young stores. Usually these pigs are older and heavier than the eight weeks used for costing purposes to distinguish between breeding and feeding stages of production for the composite breeder-feeder herds. The majority are sold at between nine and twelve weeks of age and at liveweights ranging from about 25 to 35 kg. A small number of outdoor breeding herds now sell young pigs shortly after weaning at about three to five weeks when they usually weigh under 10 kg. There were insufficient herds in the scheme selling these lightweight weaners to provide a separate sample. Table 8.1 shows average breeding results for 37 herds producing weaners and stores for sale and how they compared with the 82 herds finishing their home-bred weaners to slaughter weights.

Production standards

Although the number of pigs per litter, both born and reared, was about the same, herds selling weaners produced more litters per sow from earlier weaning, to achieve 21.0 pigs per sow a year compared with 20.3 for the breeder-feeder herds. The quantity of sow feed used per sow in herd was virtually the same for both, but those selling weaners had more pigs to share this feed to give a lower average quantity per weaner. This sow feed cost more per tonne but piglet feed was less expensive. The breeders selling weaners used more compounds, 88 per cent of the total compared with 59 per cent for herds finishing their own weaners.

Table 8.1 Average breeding results 1991

	Breeders selling weaners[a]	Breeders finishing their own weaners[b]
Number of herds	37	82
Number of sows in herd[c]	191	151
Litters per sow in herd	2.30	2.24
Age at weaning (days)	25	27
Live pigs born per litter	10.6	10.7
Weaners per litter	9.1	9.1
Weaners per sow in herd	21.0	20.3
Weight of weaners	29.4 kg	18.1 kg
Culled sow percentage	38 %	36 %
Sow feed used per sow in herd	1.30 t	1.29 t
Feed used per weaner	kg	kg
Sow feed	61.7	63.3
Piglet feed to 8 weeks	16.7	17.7
Piglet feed from 8 weeks	24.6	-
Total feed per weaner	130.0	81.0
Cost of feed per tonne	£	£
Sow feed	149.50	145.80
Piglet feed to 8 weeks	262.25	269.42
Piglet feed from 8 weeks	188.41	-
Compounds as % of total feed	88 %	59 %
Costs per weaner	£	£
Feed	18.08	13.99
Labour	6.51	5.73
Other costs	6.34	5.98
Stock depreciation	1.29	1.02
Total costs	32.22	26.72
Weaner price (net)	32.40	
Net margin per pig [d]	.18	
Capital requirements per sow	£718	£636
Return on capital	0.5 %	

(a) To sale
(b) Eight weeks of age
(c) Monthly average (including in-pig gilts)
(d) No charge has been included for interest on capital

Total costs for the herds selling weaners averaged £32.22 at sale (29.4 kg) but at eight weeks they averaged £26.51 per pig, just 21p different from the breeder-feeders at the same stage. Feed and other costs were slightly lower, while labour and stock depreciation were higher. To time of sale the heavier weaners cost more to produce, as expected, and in a year of poor prices, the average received just covered costs to leave a tiny margin (before imputing interest) of 18p each.

Long-term performance

As with all production, costs of producing weaners for sale have been increasing despite improvements in efficiency. Litter sizes increased until higher mortality reduced weaner numbers slightly in 1990 and 1991. The output of weaners per sow in herd was as usual related to the number of litters produced each year and here again there was steady improvement until 1988 (Table 8.2).

The reason for the plateau in the number of litters per sow, and subsequent weaners per sow, during and since the 1980s was because the majority of herds, having completed a change to earlier weaning, now had fewer opportunities for further improvements. The average liveweight of weaners at sale increased by about 5 kg over this 22 year period as most buyers preferred to have their pigs heavier. The extra weight produced required more piglet feed and since 1985 the total quantity per weaner has risen. This followed a long period during which the additional weaners had led to a wider division of the sow feed.

The cost of feed per tonne, in current terms, rose some fivefold between 1970 and 1991 due to change of regime through EC entry and high inflation. Higher feed costs per weaner followed the increased prices per tonne, although some containment in the quantity used did help check the rise in costs. Labour and other costs have also risen, especially in the latter years when feed prices per tonne were fairly stable. Feed costs per weaner peaked in 1984 and did not return to this level again until 1991 when they were just 1 per cent higher. In comparison, labour and other costs in 1991 were 48 per cent higher than in 1984, which highlights their increased importance in the structure of costs.

Table 8.2 Average breeding results for herds selling weaners

Year[a]	No. of herds	Litters[b] per sow in herd	Weaners[b] per sow in herd	Lwt of of weaners at sale	Feed[c] used per weaner	Cost[c] of feed per tonne
				kg	kg	£
1970	21	1.97	16.9	24	111	35.85
1971	20	2.02	17.4	25	109	43.14
1972	20	2.09	18.0	25	114	39.79
1973	24	2.09	17.8	25	108	53.98
1974	27	2.08	17.7	26	109	79.19
1975	24	2.08	18.0	25	106	79.78
1976	26	2.11	18.3	25	103	93.87
1977	28	2.10	18.4	25	98	122.08
1978	33	2.14	19.1	26	101	116.07
1979	39	2.18	19.4	26	101	128.77
1980	37	2.21	19.8	26	98	141.17
1981	37	2.27	20.0	26	97	150.32
1982	41	2.30	20.3	26	97	162.86
1983	36	2.32	21.1	26	95	170.55
1984	38	2.27	20.9	27	96	185.42
1985	35	2.33	21.6	27	95	174.84
1986	37	2.29	21.1	28	99	165.21
1987	38	2.29	21.2	28	98	165.81
1988	37	2.33	21.8	28	98	163.83
1989	31	2.28	21.1	29	102	171.32
1990	32	2.31	20.9	29	102	172.87
1991	37	2.30	21.0	29	103	176.53

(a) Year ended 30 September
(b) Includes in-pig gilts
(c) All breeding stock feed

Financial results

The overall level of profitability in weaner production depends largely on prevailing prices when sold. Prices of weaners, like those of finished pigs, frequently fluctuate and often nowadays by quite alarming amounts. When prices are high, production is usually profitable but when they are low, pigs are often produced at a loss. The annual average prices sometimes disguise short-term fluctuation but the listing in Table 8.3 showing annual rises and falls clearly indicates the good and bad years.

Table 8.3 Average costs and returns in real terms[a] for herds selling weaners

Year[b]	Feed	Labour	Other	Stock dep	Total costs	Weaner price	Margin per pig [c]	Annual total costs[d]
	£	£	£	£	£	£	£	£
1970	28.68	6.97	7.19	1.09	43.93	52.86	8.93	6.05
1971	31.17	6.74	6.94	1.47	46.32	50.73	4.41	6.94
1972	28.29	7.51	6.51	1.18	43.49	49.45	5.96	7.01
1973	33.41	8.01	7.15	.52	49.09	59.33	10.24	8.58
1974	43.45	8.16	6.96	1.65	60.22	58.72	-1.50	12.03
1975	34.78	8.18	7.57	1.07	51.60	60.27	8.67	12.61
1976	33.37	8.03	8.85	.69	50.94	66.70	15.76	14.84
1977	35.55	6.99	7.88	.97	51.39	51.89	.50	17.42
1978	31.69	7.31	8.01	1.08	48.09	57.91	9.82	17.83
1979	31.78	7.29	8.57	1.41	49.05	51.79	2.74	20.19
1980	28.23	7.17	7.99	.68	44.07	47.45	3.38	21.51
1981	27.00	6.87	8.18	1.02	43.07	44.41	1.34	23.69
1982	26.11	6.61	7.47	.58	40.77	44.83	4.06	24.68
1983	25.69	6.96	7.39	.99	41.03	37.71	-3.32	26.05
1984	26.81	6.75	6.96	.60	41.12	43.04	1.92	27.41
1985	23.61	6.55	6.84	.85	37.85	43.32	5.47	26.72
1986	22.55	6.49	6.62	.83	36.49	39.62	3.13	26.78
1987	21.45	6.38	6.55	1.00	35.38	37.58	2.20	27.00
1988	20.25	6.22	6.42	1.50	34.39	33.79	-.60	27.37
1989	20.47	6.34	5.94	.71	33.46	37.20	3.74	28.64
1990	19.01	6.54	6.59	.82	32.96	42.92	9.96	30.72
1991	18.08	6.51	6.34	1.29	32.22	32.40	.18	32.22

(a) Reflated by the retail prices index to 1991 values
(b) Year ended 30 September
(c) No charge has been included for interest on capital
(d) Annual total costs at current value

Since the end of the Guaranteed Scheme and change to the CAP from entry to the EC in 1973, weaner producers have experienced more financially bad years than good ones. Table 8.3 shows that unprofitable times occurred more or less every three or four years, while the good years of high profit were rare. The mid-1970s were quite profitable with good margins in three out of the four years from 1975 to 1978, but then 12 years passed until the next good year in 1990, although 1985 and 1989 could perhaps be classed as reasonable years.

In current terms the annual total costs at sale increased from an average of £6.05 (24 kg) in 1970 to £32.22 (29 kg) in 1991. In real terms, when costs were reflated to 1991 buying power, the total costs per weaner peaked in 1974 and have since fallen almost continuously. The average cost per weaner at sale fell from £60.22 (26 kg) in 1974 to £32.22 (29 kg) in 1991, which, allowing for weight variation, is equal to a fall of nearly 50 per cent. Feed, by far the largest item of costs, fell from £43.45 to £18.08. While labour and other costs fell during the late 1970s and early 1980s, they have remained fairly constant since the mid-1980s.

Monthly prices

The timing of the frequent price fluctuations has considerable influence on annual prices and margins. For example, the monthly average prices of weaner sales, given in Figure 8.1, show that the 1991 recording year (ending 30 September) contained two periods of low prices when weaners were under £30. Had that year run from May 1991 to April 1992, it would have included two high-priced periods and produced an average weaner price £2.50 higher than that recorded for the year which ended seven months earlier.

The weaner prices in Figure 8.1 have been standardised to 30 kg to provide a clearer illustration of price movements. Actual average weights at sale over this period were by year 0.6 to 2.0 kg per weaner below this. Although prices were previously lower for a few months in 1983, they fell again in October 1988 to £25.58, then increased nearly every month to peak at £42.77 in the following October. A price change of this magnitude in such a short time was a new experience but was soon to be repeated. At the start of 1990 prices briefly fell back to about £37 and then quickly recovered to around £43, where they remained for five months before collapsing to £28 by the end of the year. A further rise in 1991 saw prices reach £39 in May but by August they had fallen back again to £28. In 1992, prices had once more risen and fallen again, though this time not to the previous low levels.

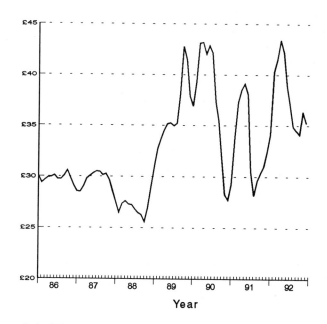

Figure 8.1 Monthly average net price per weaner sold 30 kg

Monthly margins

For most of the five years from 1988 to 1992 weaner prices were continually fluctuating while average production costs were rising steadily. Consequently profit margins varied in accordance with prices from surpluses of over £10 per pig to losses of about £5 (Figure 8.2).

The most profitable time for weaner producers was from the autumn of 1989 to the summer of 1990, when demand for pigs was strong and margins of up to £12 a head were made. Prices of 30 kg weaners for most of the time averaged around £43 a head. The least profitable times were at the end of 1990 and summer of 1991, although in between a short recovery in prices raised average margins from a loss of £5 to a surplus of £5 and back again to a loss of £5, all within a period of some eight months! While producers were pleased when prices were high and margins good, the frequent collapse in prices and subsequent losses caused many difficulties. Most would have preferred more reasonable stable prices and margins throughout.

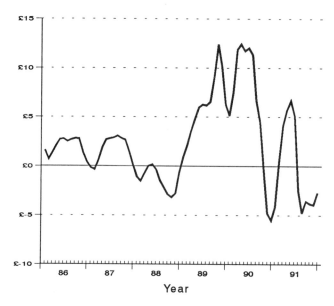

Figure 8.2 Monthly average net margin per weaner 30 kg

Range of financial results

Average results conceal the wide variations between individual herds. Examples of the range, as indicated by the 10 most profitable and 10 least profitable herds breeding herds selling weaners in the scheme in 1991 are given in Table 8.4.

Output per sow for the 10 most profitable breeding herds selling weaners was £701 for 1991, compared with £660 average of all herds and £578 for the 10 least profitable herds. The output directly reflects the number of weaners produced per sow,

Table 8.4 Financial results for breeding herds selling weaners 1991

	Average	Most[a] profitable	Least[a] profitable
Number of herds	37	10	10
Output per sow in herd	£660	£701	£578
Costs & margin per £100 output	£	£	£
Feed	57.95	53.39	69.05
Labour	20.78	18.11	25.63
Other costs	20.23	17.49	24.63
Total costs	98.96	88.99	119.31
Margin[b]	1.04	11.01	-19.31
	100	100	100
Costs and returns per pig	£	£	£
Feed	18.08	16.55	20.95
Labour	6.51	5.65	7.85
Other costs			
Farm transport	.61	.56	1.24
Vet and vet supplies	.81	.59	1.14
AI	.08	.09	.07
Power and water	.98	.73	.84
Miscellaneous	.81	.76	.84
Litter	.34	.19	.44
Maintenance	.62	.60	.46
Equipment charge	.34	.33	.49
Buildings charge	1.59	1.41	1.73
Pasture charge	.16	.19	.30
Total other costs	[6.34]	[5.45]	[7.55]
Stock depreciation	1.29	1.18	1.20
Total costs	32.22	28.83	37.55
Weaner price (net)	32.40	32.24	31.08
Margin[c]	.18	3.41	-6.47
Weight of weaners at sale	29.4 kg	30.4 kg	28.7 kg
Margin 1990	£9.28	£13.39	£5.15
1989	£3.20	£6.38	-£0.73
1988	-£0.48	£1.94	-£5.45

(a) Selected on margins per £100 output
(b) No charge has been included for interest on capital
(c) No charge has been included for interest on capital, or allowance for changing values of pigs on hand between opening and closing valuations

22.4 for the most profitable herds, compared with the average of 21.0 and 18.7 for the least profitable (Table 8.5). In a year of low prices, margins suffered but the most profitable herds achieved an average surplus of £11.01 per £100 output, while the least profitable lost £19.31.

There was little difference in the price received for weaners sold, after allowing for the difference in weight at sale, so most of the variation in profit was due to production costs, which were considerably lower for the most profitable herds. In fact, the most profitable group received a slightly poorer price (£32.24) than the average of all herds (£32.40) for marginally heavier weaners, 30.4 kg compared with 29.4 kg.

Feed costs for the 10 most profitable herds amounted to £16.55 per weaner, £1.53 less than average and £4.40 less than the 10 least profitable, which came to £20.95 each. Labour costs followed the same order with the most profitable at £5.65 per weaner, 86p less than average and £2.20 less than the least profitable. The most notable differences in other costs were for farm transport, veterinary expenses and litter. In total, other costs for the most profitable herds were 89p less than average and £2.10 lower than the least profitable. The higher buildings and equipment charges per pig for the least profitable herds were mainly a reflection of fewer weaners to share these charges.

Total costs for the most profitable herds came to £28.83 per 30.4 kg weaner, compared with the average £32.22 for 29.4 kg and £37.55 per 28.7 kg weaner for the least profitable. These costs, when deducted from the net price received for weaners sold, left margins of £3.41, £0.18 and -£6.47 per weaner respectively. Margins were much better for all three groups of herds in 1989 and 1990 because of higher prices, but 1988 was a poor year, worse than 1991 for all except the least profitable herds.

Range of production results

The most profitable herds averaging 248 sows per herd were considerably larger than the least profitable with 95 sows (Table 8.5). This probably accounts for some of the difference in cost of feed per tonne between the best and worst groups, through advantages from bulk buying, though the latter spent more on feed additives. The most profitable produced more litters per sow in herd during the year, from slightly earlier weaning. Litter size,

Table 8.5 Production results for breeding herds selling weaners 1991

	Average	Most[a] profitable	Least[a] profitable
Number of herds	37	10	10
Number of sows in herd[b]	191	248	95
Litters per sow in herd	2.30	2.35	2.19
Age at weaning (days)	25	24	27
Live pigs born per litter	10.6	10.8	10.2
Weaners per litter	9.1	9.5	8.6
Weaners per sow in herd	21.0	22.4	18.7
Sow feed used per sow	1.30 t	1.30 t	1.36 t
Feed used per weaner to sale	kg	kg	kg
Sow feed	61.7	58.1	72.4
Piglet feed to 8 weeks	16.7	17.1	17.5
Piglet feed from 8 weeks	24.6	23.5	23.4
Total feed	103.0	98.7	113.3
Cost of sow feed per tonne	£149.50	£143.99	£156.92
Cost of piglet feed per tonne to 8 wks	£262.25	£233.42	£289.14
Cost of piglet feed per tonne from 8 wks	£188.41	£179.57	£191.35
Compounds as % of total feed	88 %	90 %	90 %
Capital requirements per sow	£	£	£
Value of sow	134	133	133
Share of boars value	18	18	24
Buildings and equipment	415	411	452
Working capital	151	141	164
Total capital	718	703	773
Return on capital - 1991	0.5 %	10.8 %	-15.6 %
1990	29.0 %	43.1 %	14.1 %
1989	11.2 %	27.1 %	-2.0 %
1988	-1.7 %	7.8 %	-14.7 %

(a) Selected on margins per £100 output
(b) Monthly average (including in-pig gilts)

both born and reared, was best for the most profitable herds and with more litters produced a greater number of weaners per sow. Sow feed used per sow in herd was the same as average, but less than the least profitable herds, and with more weaners to share this feed the quantity was only 58.1 kg per weaner compared with 61.7 kg for average and 72.4 kg for the least profitable herds. The amount of piglet feed used to time of sale was only marginally different between the two groups. Overall breeding stock feed requirements, therefore, largely reflected the variation in sow feed per weaner, with the total for the most profitable herds averaging 98.7 kg for producing 30.4 kg pigs compared with 113.3 kg for the least profitable producing slightly smaller weaners of 28.7 kg. The cost per tonne for each of the three types of feed was lower for the most profitable herds.

Capital requirements included breeding stock at the average of opening and closing valuations. The value of boars was shared by the number of sows in herd. Buildings were assessed at current (replacement) values in relation to condition and equipment at cost less depreciation, apart from tractors where current values applied. Working capital covered feed, labour and other costs and was intended to reflect the average amount required for the established herds in the scheme. The total amount of capital required to start a new herd on a "greenfield" site at present day prices would be much higher than the figure in Table 8.5. On the other hand, some older buildings used by herds in the scheme which may have been written off in the farm accounts were charged at current values.

As with margins per £100 output and per pig, return on capital for 1991 was low. Only the most profitable herds achieved a reasonable return and this would have disappeared if interest had been charged on all of the assessed capital. By contrast, the previous year produced good returns all round; even the least profitable herds made 14.1 per cent on capital.

Although the variation in prices received between herds was quite small in comparison with costs, some did achieve better prices than others. Figure 8.3 shows the scatter of individual average net prices per pig sold related to their liveweight in 1991.

These are prices achieved by individual herds that sold pigs regularly throughout the recording year of 1991. Others that sold occasional batches have been omitted because of distortion from the fluctuations that occurred during the year. With the average of £32.40 for 29.4 kg the range extended from a low of £27.98 (30.1 kg) to a high of £36.83 (35.4 kg).

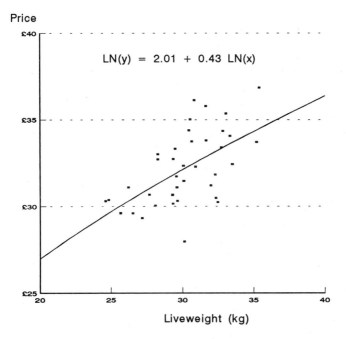

Figure 8.3 is labelled:

Price

LN(y) = 2.01 + 0.43 LN(x)

Liveweight (kg)

Figure 8.3 Net price per pig related to liveweight of weaners and stores sold 1991

Chapter 9

Finishing and Marketing
for Breeding and Feeding Herds

Type of production

Nowadays three main types of finished pigs are produced: porkers, cutters and baconers. In the latter years of the Cambridge scheme, pigs under 60 kg deadweight (approx 80 kg liveweight) were classed as porkers and those over 60 kg as cutters. Herd averages showed that most porkers were between 50 and 58 kg and cutters between 63 and 70 kg. Until the 1980s many pigs were slaughtered at lighter weights, but more recently, the demand and pricing structure have encouraged heavier weights. Previously the division between porkers and cutters was fixed at 50 kg and moved up in stages as slaughter weights increased. Some 90 per cent of porkers and cutters were sold by deadweight, mostly on contract, and were usually graded for quality. The remainder were sold liveweight in auction markets. Baconers weighed between 65 and 75 kg deadweight, the upper weight increasing in later years, and were usually sold on bacon pig contracts. In recent years a few large scale buyers ceased distinguishing between cutters and baconers with producers and just bought them as finished pigs; their destination was decided at the abattoir according to requirements and suitability.

A fourth type previously produced in quantity was heavy pigs, usually between 75 and 90 kg deadweight and used mainly for manufactured products (sausages, pies, etc.), with some parts for bacon and fresh pork. Heavy pigs were very popular in the 1960s and 1970s but suffered from high production costs as feed prices escalated. During the 1980s, the main processor of heavy pigs greatly reduced numbers and eventually stopped buying direct from producers. Since then only a small number of slaughterings have been in this category.

Each year the herds in the scheme were selected to represent the three types of production. Since 1982 there have been insufficient heavy pig producers to form the fourth group. Only herds with 80 per cent or more by value of one type of production were included and, in addition, the average weight of all pigs produced for each herd had to be appropriate for that category. For 1991, there were 25 herds in the porkers group, 47 in cutters and 29 in baconers. On average each group produced just over 3,000 pigs per herd. Herds with mixed production were omitted from these comparisons.

Finishing home-bred weaners

Average results for 1991 are shown in Table 9.1 for the herds using home-bred weaners.

The performance factors and production costs show a similar pattern of results achieved in the previous years. Poor margins were due to low finished pig prices in 1991. Prices in 1990 were some 19p per kg deadweight higher and margins per pig were the best on record, which clearly illustrates that the overall level of profitability is closely related to prevailing pig prices.

As expected, the feed conversion rate increased with the weight of pig produced, ranging from 2.50 for porkers to 2.69 for baconers. The worsening conversion rates were virtually compensated for by the fall in feed costs per tonne as the pigs became heavier. The pork group used more compounds, while both cutters and baconers included herds using by-products which lowered the price of feed per tonne. The combination of conversion rate and cost per tonne resulted in a very similar feed cost per kg liveweight gain (39.5p to 39.8p) for all groups. Labour costs per kg declined marginally as the weight of pigs increased but other costs rose. Despite varying mortality rates the charges per kg were similar for each group due to increased weight gain by the heavier pigs to share this charge. In total, costs per kg liveweight gain were virtually the same for each group.

Table 9.1 Average results by type of production 1991[a]
(breeding and feeding)

	Porkers	Cutters	Baconers
Deadweight per pig	55 kg	66 kg	69 kg
Net price per kg dwt (excl bonus)	101.0 p	97.9 p	98.4 p
Daily liveweight gain	.62 kg	.63 kg	.62 kg
Mortality percentage	2.3 %	2.8 %	2.5 %
Feed conversion rate[b]	2.50	2.63	2.69
Cost of meal per tonne[b]	£158.69	£153.83	£153.67
Cost of feed per tonne[b]	£158.69	£150.14	£147.80
Compounds as % of total feed	46 %	39 %	27 %
Costs per kg liveweight gain	p	p	p
Feed	39.7	39.5	39.8
Labour	5.3	5.2	5.1
Other costs	5.8	6.0	6.2
Mortality charge	1.1	1.1	1.0
Total	51.9	51.8	52.1
Feeding costs per pig	£	£	£
Feed	22.74	27.79	29.55
Labour	3.03	3.65	3.78
Other costs	3.32	4.22	4.60
Mortality charge	.61	.75	.67
Total feeding costs	29.70	36.41	38.60
Weaner cost (home-bred 18.1 kg)[c]	26.72	26.72	26.72
Breeding and feeding costs	56.42	63.13	65.32
Net price (including bonus)	55.79	64.78	68.46
Margin[d]	-.63	1.65	3.14
Margin 1990	£12.11	£17.09	£17.70
1989	£5.78	£8.40	£9.03
1988	-£1.86	£0.20	£0.41
1987	£3.61	£4.84	£6.43

(a) To qualify for inclusion in one of the three groups at least 80 per cent by value of pigs sold had to be in that category
(b) Includes other feed (mainly by-products) "converted" to meal equivalent (Appendix 4)
(c) Weaner charged at average production cost of the breeding and feeding herds
(d) No charge has been included for interest on capital

Costs per pig

On a per pig basis, feeding costs rise, of course, with the weight of pig. To arrive at full costs the weaner must be included. The average cost per weaner for the breeder-feeder herds was used for all groups on the assumption that whatever the type of finishing, all start with a weaner of similar standard. Details of the costs per weaner for the breeder-feeder herds are given in Table 9.2.

**Table 9.2 Average costs per weaner at eight weeks 1991
(breeder-feeder herds)**

	£
Feed	13.99
Labour	5.73
Other costs	5.98
Stock depreciation	1.02
Total costs	26.72
Litters per sow in herd	2.24
Live pigs born per litter	10.7
Weaners per litter	9.1
Weaners per sow in herd	20.3
Weight of weaners at 8 weeks	18.1 kg
Sow feed per weaner	63.3 kg
Piglet feed per weaner	17.7 kg
Cost of sow meal per tonne	£145.80
Cost of piglet meal per tonne	£269.42

The performance and costs for herds finishing their home-bred weaners are usually similar at the eight week stage to those for all herds (Table 6.5), which included some selling weaners. Marginally fewer weaners were produced per sow for herds finishing their home-bred weaners, 20.3 compared to 20.6 per sow a year for all herds. Altogether, the difference in total costs was only 8p per weaner, an insignificant amount but the appropriate figure to use for the breeder-feeder herds.

Margins per pig

The average price received per pig included any contract bonus payments but was net of haulage and other marketing charges. Baconers made the best margin per pig (Table 9.1), while porkers lost money. Cutters with a small surplus were about mid-way between the two. It should be remembered that these were average results and individual herds vary considerably. Often the best cutter producers are as profitable as bacon pig producers but as a group the latter, on average, usually achieved the best margins and return on capital, though the difference between them has narrowed recently. While some pork producers have done well, most have suffered from declining price differentials in relation to cutters and baconers in the past few years.

Other costs

Details of "other costs" per pig for the feeding period are given in Table 9.3. They are also shown for breeding and feeding together.

Charges for buildings and equipment for feeding stock form approximately 44 per cent of other costs for porkers, 40 per cent for cutters and 36 per cent for baconers. Bacon producers appear to spend more per pig on veterinary costs while pork producers spend less on most items, probably because their pigs are smaller and spend less time on the farm. Unexpectedly the buildings charges and costs of power and water were marginally lower for baconers than for cutters who on average have a slightly shorter finishing period.

For breeding and feeding combined, the buildings and equipment charges were around 35 per cent of other costs in all groups. Between 1988 and 1991 these other costs have increased by over a third per pig. All items show substantial increases apart from buildings and equipment, though these have also risen but at a lower rate. Labour costs also increased over this period by some 23 to 32 per cent, but feed was only a few pence higher. Total costs provide further evidence of the change in the cost structure of pig production, with labour and other costs now more important than before, accounting on average for 32 per cent of costs in 1991 compared with 27 per cent in 1988.

Table 9.3 Average other costs per pig 1991
(breeding and feeding)

	Porkers	Cutters	Baconers
Feeding from 8 weeks	£	£	£
Farm transport	.32	.48	.44
Vet and vet supplies	.16	.19	.27
Power and water	.28	.53	.47
Miscellaneous expenses	.45	.58	.56
Litter	.30	.31	.55
Maintenance	.36	.44	.67
Equipment charge	.20	.32	.34
Buildings charge	1.25	1.37	1.30
Total other costs	3.32	4.22	4.60
Breeding and feeding combined	£	£	£
Farm transport	.68	.84	.80
Vet and vet supplies	1.01	1.04	1.12
AI fees	.12	.12	.12
Power and water	1.53	1.78	1.72
Miscellaneous expenses	1.03	1.16	1.14
Litter	.62	.63	.87
Maintenance	.95	1.03	1.26
Equipment charge	.48	.60	.62
Buildings charge	2.84	2.96	2.89
Pasture charge	.04	.04	.04
Total other costs	9.30	10.20	10.58

Costs per kg deadweight

Another basis of comparison is by production costs per kg deadweight. Finishing pigs are sold by weight, over 90 per cent by deadweight, and costs shown this way (Table 9.4) are often helpful in selecting suitable outlets for pigs and in explaining how profitability relates to these costs. Here feeding costs are given separately from weaner costs and in relation to the total deadweight per pig.

**Table 9.4 Average costs and margin per kg deadweight 1991
(finishing home-bred weaners)**

	Porkers	Cutters	Baconers
Feeding costs	p	p	p
Feed	41.3	42.1	42.6
Labour	5.5	5.5	5.5
Other costs	6.0	6.4	6.6
Mortality charge	1.1	1.1	1.0
Total feeding costs	53.9	55.2	55.7
Weaner costs (home-bred)	48.5	40.5	38.5
Total costs	102.4	95.7	94.2
Net price (incl bonus)	101.3	98.2	98.7
Margin per kg	-1.1	2.5	4.5
Total costs per kg 1990	98.6	91.4	91.5
1989	96.9	91.6	90.5
1988	94.3	87.1	86.9
Net price per kg 1990	120.5	117.1	117.5
1989	107.5	104.4	103.8
1988	90.8	87.4	87.4

Feed costs slightly more per kg deadweight as the pigs get heavier (from 41.3p to 42.6p in 1991) due to worsening conversion rates. Labour costs were the same in each case but other costs also increased. In total, feeding costs per kg rise as the pigs increase weight but when the weaner is included, the larger pigs have a definite advantage. The cost of the weaner is fixed whatever the size of pig produced. It follows, therefore, that the larger the pig the lower the cost of weaner per kg deadweight as it is divided by more weight. The weaner cost per kg falls sharply from 48.5p for porkers down to 38.5p for baconers. This fall more than compensates for the small rise in feeding costs and total costs drop from 102.4p per kg for porkers to 94.2p for baconers.

Although costs of production per kg decline as the pig increases in size, the receipts follow the same pattern except for the more specialised bacon pigs. In their heyday heavy pigs of up to 90 kg always returned a lower price per kg than cutters. This still applies to the few smaller heavy pigs currently produced. The low pig prices of 1991 failed to cover average costs for porkers but left small surplus margins per kg for cutters and baconers. Total costs and prices received per kg are also given for the three previous years and show that while costs have in general been increasing steadily, pig prices recovered in 1989 and 1990, only to fall again in 1991.

Monthly pig prices

Annual averages tend to smooth out repetitive fluctuations of pig prices. Every few months since 1988 price changes have been

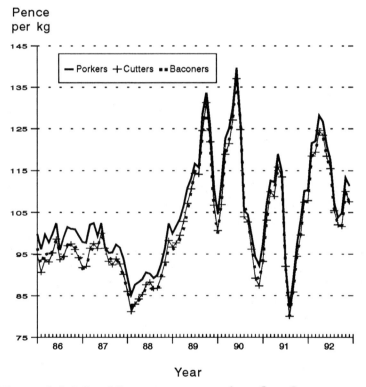

Figure 9.1 Monthly average net price of porkers, cutters and baconers per kg deadweight

greater and more frequent than ever before, as shown in Figure 9.1, where average prices for porkers, cutters and baconers are plotted on a monthly basis. Previously, when prices had risen steadily from the early 1970s to 1984, there had been fluctuations but these were minor compared to those of recent years. Prices had fallen in 1988 but then rose to new record heights in 1989 and 1990. Further peaks were registered in 1991 and in 1992, after the scheme had finished. In between times prices had collapsed on four occasions. The most crucial falls were in 1990 and 1991 with less serious lows at the end of 1989 and in 1992. For the last five years of the graph, it can be seen that pig prices changed often and quite rapidly by substantial amounts. At times like this, budgeting was made very difficult for pig producers. It seems unlikely that anything can be done to prevent such chaotic situations recurring in the future.

The graph also shows that the price differential between cutters and baconers almost disappeared in later years, as the lines often appear to merge. With continuing changes taking place, it is unclear whether pork prices have fallen, or whether cutter and bacon prices have risen to close the gap. A likely guess would be the former!

Monthly margins per pig

As pig prices fluctuated frequently so have profit margins. Costs have been rising at a gradual rate from higher labour and other costs, while feed costs have been relatively stable. These costs, however, vary seasonally, as performance in the summer months is superior to the winter, when more pigs are produced per sow and feed requirements are less. Feed prices are usually at their lowest at harvest and increase with time until the next harvest. Margins per pig on a monthly basis appear in the following graphs; they are given separately for porkers, cutters and baconers. Margins are given for breeder-feeders and also for feeding-only herds, which are dealt with in more detail in the following chapter.

For all types of production, margins for breeding and feeding herds were usually better than for feeding-only herds relying on purchased weaners. Normally the price of purchased weaners includes a profit margin for the breeder and cost more than home-bred weaners. At times of low demand weaners can sometimes be bought at below cost of production; it is then that feeding-only is likely to be more profitable than breeding and feeding combined. When weaners are cheap, the feeding-only herds have an advantage but they lose out when weaners are expensive. Weaners purchased from several different sources present potential health hazards which may affect performance and production costs.

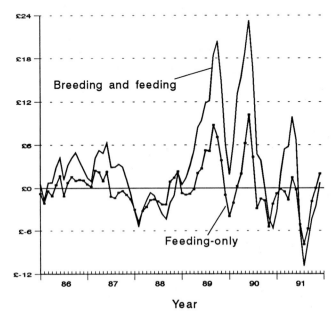

Figure 9.2 Monthly average net margins per pig for producing porkers (interest not charged)

The monthly margins for herds producing porkers, shown in Figure 9.2, clearly illustrate the extent of the fluctuations in profitability. For breeding and feeding average margins improved steadily over a 20 month period from -£5.44 per pig in 1988 to +£20.42 in October 1989. Since then rapid changes every few months brought low margins in January 1990, December 1990 and August 1991 and high margins in June 1990 and May 1991. The best margin over this period was £23.29 per pig (June 1990) and the worst minus £10.90 (August 1991), a difference of over £34 in just 14 months, with a trough and peak in between. The timing was similar for feeding-only herds, though occasionally margins changed direction sooner following earlier changes in weaner prices. Here the best margin was £10.23 (June 1990) and the worst minus £7.89 (August 1991). There were three periods in 1988, 1990 and 1991 when feeding-only herds did better than breeding and feeding. These were all unprofitable times when feeders had the benefit of comparatively inexpensive weaners.

The fluctuations in profitability occurred at the same times for cutters and baconers, but usually the margins were better than for porkers.

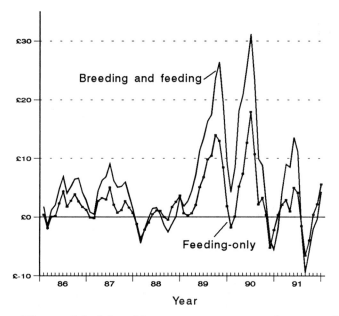

Figure 9.3 Monthly average net margins per pig for producing cutters (interest not charged)

For cutter production the best margins achieved were £31.14 per pig for breeder-feeders and £17.95 for feeding-only, both in June 1990. The worst were -£9.45 and -£6.54 respectively in August 1991, as shown in Figure 9.3.

The best margins for bacon pig producers were also in June 1990, when breeder-feeders averaged £31.59 per pig and feeding-only £18.25. Corresponding lows in the following August were -£8.46 and -£5.61 respectively for the two systems; see Figure 9.4.

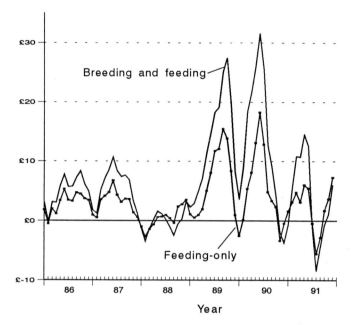

Figure 9.4 Monthly average net margins per pig for producing baconers (interest not charged)

The profitability of all types of pig production has frequently changed since 1988 from sizable losses to substantial surpluses. Moreover these changes have occurred very quickly. In a little under two years, between October 1989 and August 1991, margins have swung drastically on five occasions at intervals of three to six months. In just three months the high margins of October 1989 had fallen by up to £23 per pig, then rose by £28 in the following five months, only to crash by £36 in the next six months, then recover again by £19 in five months and fall once more by £23 three months later. Few in the pig industry could offer producers a satisfactory explanation for this state of affairs. Seasonal factors such as hot weather no longer appear to influence pig prices, as 1990, 1991 and 1992 all had the highest prices of the year in May or June, though in 1991 August prices were at their lowest for many years.

Variation in pig prices

As most pig producers sell their pigs fairly evenly throughout the year, annual average prices should not be unduly affected by short-term fluctuations, unless the timing allows more peaks or troughs in one year than another. Anyone selling pigs only three or four times a year may receive high prices more often than low, or vice versa, but prices usually even out over the year for regular suppliers, so that individual herd prices do not differ greatly from the overall average.

All herds included in Figure 9.5, showing the relation of net pig prices per kg to deadweight of pigs sold in 1991, marketed their pigs regularly. Some producers sold pigs in more than one category.

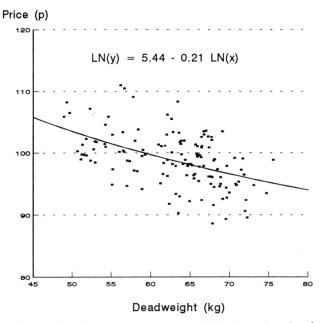

Figure 9.5 Net price per kg related to deadweight of porkers, cutters and baconers sold 1991

Table 9.5 Financial results for breeding and feeding herds 1991

Breeding and feeding	Average	Most[a] profitable	Least[a] profitable
Number of herds	82	20	20
Output per sow in herd	£1,207	£1,314	£1,026
Costs & margins per £100 output	£	£	£
Feed	65.89	60.36	73.51
Labour	15.24	13.46	19.40
Other costs	16.33	14.30	19.25
Total costs	97.46	88.12	112.16
Margin[b]	2.54	11.88	- 12.16
	100	100	100
Costs and returns per pig[c]	£	£	£
Feed	40.83	37.85	44.53
Labour	9.38	8.42	11.43
Other costs			
Farm transport	.76	.83	.89
Vet and vet supplies	1.03	1.04	1.33
AI	.13	.11	.10
Power and water	1.75	1.49	2.15
Miscellaneous expenses	1.15	1.10	1.53
Litter	.67	.70	.66
Maintenance	1.07	.82	1.05
Equipment charge	.55	.48	.55
Buildings charge	2.89	2.37	3.05
Pasture charge	.04	-	.04
Total other costs	[10.04]	[8.94]	[11.35]
Stock depreciation	1.05	.72	1.69
Total costs	61.30	55.93	69.00
Net price (incl bonus)[d]	62.39	63.41	61.66
Margin[b]	1.09	7.48	-7.34
Average deadweight of sales	62.8 kg	63.9 kg	61.9 kg
Net price per kg (excl bonus)	99.0 p	98.9 p	99.4 p

(a) Selected on margins per £100 output
(b) No charge has been included for interest on capital
(c) Breeding and feeding costs combined
(d) Haulage, marketing charges and levies have been deducted

Deadweight per pig ranged from just under 50 kg to 75 kg and included porkers, cutters and baconers. Prices received ranged from 88.6p per kg (68 kg) to 111p (56 kg), with an overall average 98.7p at 64.2 kg deadweight. The trend line in the graph confirms that price per kg can be expected to fall as the weight of pig increases. Individual readings show a wide variation in prices but most were close to the trend line. Part of the variation in price was likely to be due to the grading of pigs for quality. Most of the pigs sold were graded but unfortunately buyers classified their pigs differently, making the collection of gradings cumbersome, especially for those producers selling to more than one outlet. Collecting records of gradings served no useful purpose and was not undertaken. It had to be assumed that the quality of pigs was reflected in the price they made.

Most and least profitable herds compared

Of the 82 breeding and feeding herds in the scheme in 1991, the 20 most profitable in terms of margin per £100 output were selected to compare results with the 20 least profitable. The overall margin for the 82 herds finishing their own home-bred weaners was £2.54 per £100 output in a year depressed by low pig prices. The margin for the 20 most profitable herds averaged £11.88, from an output of £1,314 per sow in herd, compared with a loss of £12.16 for the least profitable herds with a lower output of £1,026 per sow. Details of the financial results are given in Table 9.5.

Pigs produced by the 20 most profitable herds averaged 63.9 kg deadweight, 2 kg more than the least profitable, but costs per pig were lower in every item except AI and litter. Feed was £6.68 less, labour £3.01, other costs £2.41 and breeding stock depreciation 97p, a total of £13.07 per pig. They also received on average £1.75 more for their pigs, due to the extra weight, which more than compensated for a marginally lower price kg deadweight. Despite all the additional costs incurred, the least profitable herds received only a fractional better price than the most profitable with low costs. As a result margins were much better for the most profitable herds at £7.48 per pig, compared with an average of £1.09 for all herds and a loss of £7.34 for the least profitable.

Table 9.6 Production results for breeding and feeding herds 1991

Breeding and feeding	Average	Most[a] profitable	Least[a] profitable
Number of sows in herd	151	139	104
Litters per sow in herd	2.24	2.29	2.15
Age at weaning (days)	27	26	30
Live pigs born per litter	10.7	10.8	10.5
Weaners per litter (8 wks)	9.1	9.5	8.6
Weaners per sow in herd	20.3	21.6	18.6
Pigs sold per sow in herd[b]	19.8	21.1	18.1
Sow feed used per sow	1.29 t	1.27 t	1.27 t
Feed used per pig sold	kg	kg	kg
Sow feed per weaner	64.9	60.2	70.4
Piglet feed to 8 weeks	18.1	17.9	18.9
Feeding stock feed per pig[c]	<u>173.3</u>	<u>168.0</u>	<u>186.6</u>
Total feed	<u>256.3</u>	<u>246.1</u>	<u>275.9</u>
Cost of feed per tonne[d]			
Sow	£145.78	£142.90	£145.90
Piglet	£269.42	£270.58	£262.92
Feeder	£152.80	£145.38	£156.94
Compounds as % of total feed	46 %	18 %	58 %
Feeding stock[c]			
Daily liveweight gain	.63 kg	.70 kg	.59 kg
Mortality	2.5 %	2.5 %	2.7 %
Feed conversion rate	2.60	2.47	2.85

(a) Selected on margins per £100 output
(b) Feeding stock mortality deducted
(c) From eight weeks
(d) Included other feeds "converted" to meal equivalent

Production efficiency was far superior for the most profitable herds. They achieved three more pigs per sow a year from more and larger litters, at both birth and eight weeks. The quantity of feed used per sow was the same for both groups (1.27

tonnes a year) but producing more pigs to share this feed gave the most profitable herds a distinct advantage of 60 kg per pig compared with 70 kg. Less piglet feed was used by the most profitable and, with a much better feed conversion rate, the quantity per feeder was considerably lower. In total, they required 246.1 kg of feed to produce a 63.9 kg deadweight pig, whereas the least profitable herds used 275.9 kg to produce a smaller 61.9 kg pig. The production results for the breeder-feeder herds are shown in Table 9.6.

Approximately two-thirds of the £6.68 difference in feed costs per pig between the most profitable and least profitable herds was due to the quantity of feed used. The other one-third came from the feed price; for sows this was £3 per tonne and for feeders £11.56 per tonne cheaper for the most profitable, though their piglet feed was £7.66 per tonne dearer than that used by the least profitable herds. Only 18 per cent of feed for the most profitable herds was compounds, compared to 58 per cent for the least profitable. Also, the least profitable were on average smaller herds and this may have contributed to some of the difference in cost of feed per tonne, through having to buy in smaller quantities.

The comparison of financial and production results shows that the success of the most profitable herds was due to the standard of performance achieved. They produced more pigs per sow and used less feed of lower cost per tonne. Labour and other costs were also lower. As a group, the average price received for pigs sold was very similar to that for the least profitable herds and contributed little towards the considerable difference in margins per £100 output or per pig.

Chapter 10

Finishing and Marketing Purchased Weaners

Production and marketing of finished pigs for feeding-only herds follows much the same pattern as for breeding and feeding herds, except that having no breeding stock they start by buying weaners. Many feeding-only units are long established and some have been buying weaners from the same source for many years. Some supply boars or contribute towards their cost. Usually the boars are selected to enhance performance in feed conversion and carcass conformation from which the feeders benefit rather more than the breeder.

Weaner prices

The pricing arrangements for weaners may follow profit-sharing formulas or the reported going rate for the area. Prices do vary, however, often depending on quality and batch size. Most are traded direct between the breeder and the feeder, though some go through auction markets or marketing companies. Over the years many pig breeders have given up production, forcing feeders to seek supplies of weaners elsewhere. As most breeders already have an outlet for their weaners, a new buyer must usually offer a better price than the breeder was previously obtaining to secure his pigs, especially at times when weaners are in short supply. The contract pig finishing schemes buy large numbers of weaners and often fix prices according to the profitability of pig production. Weaner prices are nearly always related to finished pig prices but in good times competition for weaners is often fierce, with some large buyers apparently willing to pay high prices to increase their involvement in pigs. The same buyers are likely to reduce numbers when pigs are unprofitable, which sometimes leaves breeders seeking alternative outlets for weaners and prices may suffer. Few breeders have sufficient accommodation to house weaners for long, and the accommodation requirements of younger weaners following on can force a sale at relatively low prices. Like finished pig prices, extremes in weaner prices seldom last for more than a few months and average prices over the year are more representative. Figure 10.1 shows individual prices per pig related to liveweight for weaners and stores purchased for the feeding-only herds in 1991.

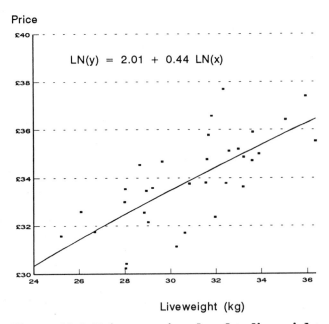

**Figure 10.1 Price per pig related to liveweight
of weaners and stores purchased 1991**

Most buyers paid for the weaner haulage and this sum was included in the prices given. Any payments to breeders for boars was shared by the number of weaners purchased during the year and included in the prices. Two herds buying small weaners of 6 kg and 10 kg have been omitted.

The trend line shows that, as expected, prices paid for weaners and stores purchased increased with the weight of pigs. Prices ranged from £32.42 for pigs averaging 28.1 kg to £37.69 for 32.3 kg and weights ranged from 25.2 kg (£31.58) to 34.4 kg (£35.53). The overall average, including the two herds buying small weaners, was £32.79 for 28 kg.

Costs and margins

An assessment of finishing purchased weaners for porkers, cutters and baconers is shown in Table 10.1. Starting with a 28 kg pig (10 kg heavier than the breeder-feeders) the feeding-only herds had less weight gain to make in reaching slaughter weight than the breeding and feeding herds. The heavier weaners gave a better daily liveweight gain and lower mortality but, as expected, a worse feed conversion rate. The cost of feed per tonne was less expensive from avoiding much of the relatively high cost starter feeds for the young pigs.

The feed conversion rates were lowest for porkers at 2.63 and increased with the weight of pig to 2.75 for cutters and 2.82 for baconers. Cost of feed per tonne was highest for porkers, followed by cutters and baconers where more other feeds (by-products) were used, which offset the advantage from the best conversion rate as all three types had a similar feed cost per kg liveweight gain of just over 40p. Labour costs per kg liveweight gain were not much different from the breeder-feeder herds but other costs were higher. The lower mortality rates for feeding-only herds, probably due to starting with older and stronger weaners, offset the higher cost of weaners. Total costs for all three types were similar at just under 54p per kg.

Feeding costs per pig increased with the weight of pig produced. For the feeding-only herds finishing purchased weaners, the average cost of weaners bought by these herds during the year (£32.79 for 28 kg) was used to arrive at total costs. The poor prices received for finished pigs in 1991 were insufficient to cover costs for porkers and cutters but left a small margin for baconers. Margins were better in 1990 due to higher prices but as a result, weaners were also much more expensive.

Margins per pig for feeding-only herds are normally expected to be less than those for the breeder-feeder herds where the production of weaners is undertaken as well as finishing

Table 10.1 Average results by type of production 1991[a]
(finishing purchased weaners)

Feeding-only	Porkers	Cutters	Baconers
Deadweight per pig	55 kg	66 kg	69 kg
Net price kg dwt (excl bonus)	101.0 p	97.9 p	98.4 p
Daily liveweight gain	.68 kg	.68 kg	.66 kg
Mortality percentage	1.6 %	2.0 %	1.8 %
Feed conversion rate[b]	2.63	2.75	2.82
Cost of meal per tonne	£154.10	£149.47	£149.46
Cost of feed per tonne[b]	£154.08	£145.78	£143.59
Costs per kg liveweight	p	p	p
Feed	40.5	40.2	40.4
Labour	5.5	5.4	5.3
Other costs	6.8	6.9	7.1
Mortality charge	1.1	1.1	.9
Total	53.9	53.6	53.7
Feeding costs per pig	£	£	£
Feed	19.14	24.23	26.00
Labour	2.60	3.26	3.41
Other costs	3.22	4.16	4.56
Mortality charge	.53	.65	.58
Total feeding costs	25.49	32.30	34.55
Weaner cost (purchased 28 kg)[c]	32.79	32.79	32.79
Total costs	58.28	65.09	67.34
Net price (including bonus)	55.79	64.78	68.46
Margin[d]	-2.49	-.31	1.12
Margin 1990	1.89	6.74	7.20
1989	2.16	4.81	5.47
1988	-2.20	-.08	.17
1987	.89	2.09	3.76

(a) To qualify for inclusion in one of the three groups at least 80 per cent by value of pigs sold had to be in that category
(b) Includes other feed (mainly by-products) "converted" to meal equivalent
(c) At average cost for all pigs purchased by feeding-only herds
(d) No charge has been included for interest on capital

pigs. The cost of weaners to the feeder includes the breeder's profit and the cost of moving them from farm to farm. The cost of transporting pigs has increased over the years and if the buying of weaners is arranged by a third party, an additional fee is incurred for this service. Occasionally, when pig prices are low, weaners change hands at below cost of production for average efficiency. The feeder then buys them cheaply and saves the money that the breeder, or breeder-feeder, lost in producing them. It is at these exceptional times that feeding-only herds will do better financially than breeder-feeders, though both could be in a loss situation.

By deadweight

Costs per kg deadweight, Table 10.2, show the advantage of taking pigs on to heavier weights. The higher costs for porker production would have been justified had the price paid for these smaller pigs been more but, on average, the 1991 price including bonus was only 3.1 p per kg more than cutters and 2.6p more than baconers. In fact, total premiums of some 7p to 9p per kg were necessary for porkers to equal the margins achieved by cutters and baconers. The decline in price differentials has been a further incentive for producers of the smaller pork pigs to raise the weights at slaughter.

Most of the feeding costs per kg deadweight for the herds finishing purchased weaners were lower than for the breeder-feeder herds. This was because the pigs were growing more quickly and less weight gain was required from starting with a purchased weaner of 28 kg than from a home-bred weaner of 18 kg. Here again, costs per kg increased with the weight of pigs. The cost per kg of bigger purchased weaners more than offset lower feeding costs, to give total costs for feeding-only herds of about 3p per kg (3.4p for porkers) higher than for breeder-feeder herds. As a result, margins were that much lower for 1991.

The benefit of spreading the overhead cost of the weaner over a larger pig is clearly demonstrated here. When shared by the 55 kg porker weight, the weaner cost is 59.5p per kg but when shared by the 66 kg cutter it falls to 49.7p per kg, nearly 10p per kg less. Weaners were more expensive in 1990 and their costs per kg were considerably higher. But this was a profitable year and prices paid for finished pigs more than covered these higher costs.

Table 10.2 Average costs and margins per kg deadweight 1991
(finishing purchased weaners)

Feeding-only	Porkers	Cutters	Baconers
Feeding costs	p	p	p
Feed	34.8	36.8	37.5
Labour	4.7	4.9	4.9
Other costs	5.8	6.3	6.6
Mortality charge	1.0	1.0	.8
Total feeding costs	46.3	49.0	49.8
Weaner cost (purchased)	59.5	49.7	47.3
Total costs	105.8	98.7	97.1
Net price (incl bonus)	101.3	98.2	98.7
Margin per kg	-4.5	-.5	1.6
Weaner cost per kg 1990	68.4	57.1	55.7
1989	56.9	47.5	45.8
1988	52.2	42.3	40.3
Total costs per kg 1990	117.1	107.0	107.0
1989	103.5	97.0	95.8
1988	95.0	87.5	87.3
Net price per kg 1990	120.5	117.1	117.5
1989	107.5	104.4	103.8
1988	90.8	87.4	87.4
Margin per kg 1990	3.4	10.1	10.5
1989	4.0	7.4	8.0
1988	-4.2	-0.1	0.1

Variation between herds

Average results conceal the wide variation that exists for individual herds. As with the breeding herds selling weaners and the breeder-feeder herds producing finished pigs, the range of results for feeding-only herds is demonstrated by comparing performance of the most profitable herds with the least profitable. The herds were selected on margins per £100 output and are shown in Table 10.3.

Table 10.3 Financial results for feeding-only herds 1991

Feeding-only	Average	Most[a] profitable	Least[a] profitable
Number of herds	31	10	10
Output per feeder place a year	£126	£135	£109
Costs & margin per £100 output	£	£	£
Feed	77.40	68.25	98.03
Labour	10.45	9.58	11.34
Other costs	13.79	13.14	15.76
Total costs	101.64	90.97	125.13
Margin[b]	-1.64	9.03	-25.13
	100	100	100
Costs and returns per pig	£	£	£
Feed	25.21	24.55	26.27
Labour	3.41	3.44	3.04
Other costs			
Farm transport	.49	.47	.48
Vet and vet supplies	.31	.36	.30
Power and water	.38	.36	.36
Miscellaneous expenses	.56	.62	.50
Litter	.45	.56	.34
Maintenance	.49	.59	.58
Equipment charge	.35	.42	.36
Buildings charge	1.46	1.34	1.30
Total other costs	[4.49]	[4.72]	[4.22]
Mortality charge	.61	.53	.75
Total feeding costs	33.72	33.24	34.28
Weaner cost (purchased)[c]	32.79	32.07	35.16
Total costs	66.51	65.31	69.44
Net price (incl bonus)[c]	66.08	68.63	62.87
Margin[b]	-.43	3.32	-6.57
Average deadweight of sales	67.2 kg	69.2 kg	65.5 kg
Net price per kg (excl bonus)	98.1 p	99.0 p	95.7 p

(a)   Selected on margins per £100 output
(b)   No charge has been included for interest on capital
(c)   Haulage, marketing charges and levies have been added to the cost of
       pigs purchased and deducted from the value of pigs sold

The pigs produced by the 10 most profitable feeding-only herds were heavier (69.2 kg deadweight) than the average (67.2 kg) and the 10 least profitable herds (65.5 kg). On average, the 31 feeding herds just failed to cover costs in 1991 and lost £1.64 per £100 output. The most profitable herds made a surplus margin of £9.03, while the least profitable lost heavily at £25.13, providing clear evidence of the range of results that exists between the best and worst herds. Output per feeder place was £135 for the most profitable compared with £109 for the least profitable, and was achieved by producing more valuable pigs at sale from smaller, less costly weaners brought in, giving an added value of £36.56 (£68.63 - £32.07) per pig against £28.59 (£62.87 - £35.16) per pig. The least profitable herds not only sold their pigs at lighter weights; they also received a lower price per kg deadweight.

Despite a greater weight gain from producing heavier pigs from smaller weaners, the most profitable herds managed a lower feed cost per pig than the average or least profitable herds and this partly disguised their low cost of production. Labour and other costs, however, were higher per pig partly because of the extra weight gain required.

Feeding costs amounted to £33.24 per pig for most profitable herds compared to £34.28 for the least profitable but when weaner costs were added, the total per pig came to £65.31 and £69.44 respectively, a difference of some £4, for producing pigs of 69.2 kg and 65.5 kg. The net price of £68.63 left a surplus of £3.32 per pig for the most profitable, while the £62.87 received by the least profitable failed to cover costs and incurred a deficit of £6.57 per pig.

The most profitable feeding-only herds were larger averaging 1,362 pigs per herd, nearly twice as many as the least profitable with 727 pigs per herd. With 5,208 pigs produced per herd, turnover for the most profitable equalled 3.8 times a year, whereas the least profitable producing 3,021 smaller pigs had a turnover of 4.1 times. The former brought pigs in at an average of 25.4 kg and reared them to 92.5 kg, a liveweight gain of 67.1 kg per pig, while the latter group brought pigs in heavier at 34 kg and sold them lighter 87.8 kg, to give a lower liveweight gain of 53.8 kg. This enabled the least profitable herds to have a better turnover rate. The liveweight gain per pig, when divided by the daily liveweight gain, indicate that the pigs from the most profitable herds spent on average 97 days on the farm, compared with 90 for the least profitable.

Table 10.4 Production results for feeding-only herds 1991

	Average	Most[a] profitable	Least[a] profitable
Number of herds	31	10	10
Number of feeders per herd	931	1,362	727
Number of pigs produced per herd	3,695	5,208	3,021
Liveweight of pigs produced	89.7 kg	92.5 kg	87.8 kg
Liveweight of pigs brought in	28.0 kg	25.4 kg	34.0 kg
Daily liveweight gain	.65 kg	.69 kg	.60 kg
Mortality percentage	1.8 %	1.6 %	2.1 %
Feed conversion rate[b]	2.87	2.83	3.01
Cost of meal per tonne	£150.55	£143.04	£162.72
Cost of feed per tonne[b]	£142.24	£129.93	£162.16
Compounds as % of total feed	30 %	7 %	78 %
Costs per kg liveweight gain	p	p	p
Feed	40.8	36.8	48.8
Labour	5.5	5.1	5.6
Other costs	7.2	7.1	7.8
Mortality charge	1.0	.8	1.4
Total feeding costs	54.5	49.8	63.6

(a) Selected on margins per £100 output
(b) Includes other feeds "converted" to meal equivalent

Performance factors shown in Table 10.4 were all better for the most profitable herds. The superiority of the most profitable feed conversion was greater than at first appears because of the larger weight of pigs produced. By far the major difference between the two groups, however, was the cost of feed per tonne. The most profitable herds used less compounds and more inexpensive other feeds (by-products) to achieve average feed costs of £32 per tonne less than the least profitable herds. Feed costs per kg liveweight gain, therefore, were much lower at 36.8p per kg compared to 48.8p. The difference was further extended through lower labour and other costs to give total costs of 49.8p for the most profitable and 63.6p for the least profitable.

Chapter 11

Systems Compared

Margins by system

There are three distinct systems of pig keeping:

1. Combined breeding and feeding
2. Mainly breeding weaners for sale
3. Feeding-only from purchased weaners

The profitability of breeding herds selling weaners and feeding-only herds is usually related to the price of weaners. When they are cheap, the feeders have the advantage and when they are expensive, the breeder benefits. The farmer who breeds and finishes his own pigs is, of course, isolated from weaner prices and his profits are steadier for that reason. He also avoids the costs involved in buying and selling weaners, that is haulage and procurement charges.

As expected margins for the three systems usually move in harmony with each other but occasionally at different rates. Changes in costs, say because of rapidly rising feed prices, have greater immediate impact on feeding-only herds with their shorter production cycle of some three to four months, compared with six to seven months for breeding herds selling weaners and nine to ten months for combined breeding and feeding herds. Weaner prices usually reflect finished pig prices, so that when they change, income from sales affects the three systems simultaneously. As feeding-only herds buy in as well as sell pigs, they stand to lose more to start with from having bought expensive weaners a few months earlier and then selling them cheaply when prices fall. They gain, however, when the reverse happens and prices rise. On a falling market, feeding-only herds suffer a bigger immediate loss than others but then recover more quickly as the replacement weaners are taken in at the new lower price.

Average margins for the three systems are shown in Table 11.1.

Table 11.1 Average margins per £100 livestock output by system

	Breeding & feeding	Mainly breeding	Feeding-only
	£	£	£
1970	17.06	15.96	13.61
1971	10.73	7.23	12.07
1972	16.96	13.69	16.85
1973	20.46	19.66	20.17
1974	2.43	-0.65	6.00
1975	22.01	19.87	23.79
1976	16.52	25.62	9.55
1977	0.16	0.95	3.20
1978	15.40	18.97	12.10
1979	3.58	5.88	1.62
1980	9.24	9.40	12.06
1981	7.53	4.08	8.10
1982	9.69	10.00	9.43
1983	-6.30	-10.44	1.46
1984	10.37	8.45	13.21
1985	10.90	10.75	8.52
1986	7.41	8.36	7.91
1987	8.33	5.87	5.57
1988	-1.57	-3.72	-1.11
1989	19.73	17.41	26.06
1990	17.37	21.12	13.15
1991	2.54	1.04	-1.64

Over a period of time margins usually even out. For the five years 1987 to 1991, the average margin per £100 output was £9.28 for breeder-feeders, £8.34 for mainly breeding and £8.41 for feeding-only. But for the problems with Aujeszky's disease which affected feeding-only herds less than others, the previous five years would also have produced similar averages.

Marketing outlets compared

A detailed supplementary study of the herds recording in the scheme was undertaken in 1986 to compare prices received from various marketing outlets. The results were published in that year's report and are briefly summarised here (Table 11.2). There were four main outlets. Direct sales covered producers marketing their own pigs direct to the buyers. Producer groups and marketing companies were two categories of intermediaries organising sales to buyers on behalf of the producers. Auction markets were for liveweight sales where arrangements were usually made by producers. A fifth category was for sows sold to dealers. The prices achieved by each outlet were related to the overall average for that class of sale and shown as better (+) or worse (-) than average. Prices for weaners and stores were adjusted to standard weights.

Sows sold through producers' groups and auction markets achieved the best prices but, as no weights were recorded, there was no way of telling whether this was simply because they were heavier. Auction markets had 34 per cent of the total sales and producers' groups 11 per cent. Direct sales with 29 per cent of the sows made below average prices.

Direct sales accounted for about half of all weaners and stores sold. They received better than average prices in each category. The marketing companies, with 27 per cent of sales, also received better than average prices for stores. Auction markets had the lowest prices that year for both weaners and stores but numbers involved were very small, only 1 and 2 per cent of the total.

Auction markets, however, were clearly the best on price for porkers and cutters (when prices were converted to a deadweight basis for comparison). Here again, the numbers were small with 10 per cent of porkers and 2 per cent of cutters. They were also lighter pigs by between 4 and 9 kilograms than those sold by other outlets.

Baconers were pigs sold for this purpose, usually on contract, regardless of whether they were actually used for bacon. Most were marketed by direct sales (53 per cent) and by producer groups (30 per cent) at very similar prices.

In choosing an outlet for their pigs, producers should consider two other important factors in addition to the price they receive. One is the duration of the period before payment is received and two is the risk of the buyer being unable to pay. In the past pigs were usually paid for promptly but nowadays some producers have to wait for three weeks. Those selling pigs every week will always have a sizable sum owing them. Although it is unusual for a pig buyer to cease trading unexpectedly, any producers unfortunate enough not to be paid for their pigs could find the experience costly. Some groups now run insurance schemes as a safeguard to cover producers' losses in such cases and these are worthy of consideration.

Table 11.2 A comparison of prices by marketing outlet, 1986

	Sows[(a)]		Weaners[(b)]		Stores[(b)]	
Net price per pig related to						
average of all herds (+ or -)	£	%	£	%	£	%
Direct sales	-2.47	(29)	+.46	(48)	+.12	(56)
Producers' group	+2.62	(11)	-.20	(32)	-.17	(13)
Marketing companies	-3.48	(8)	-.82	(12)	+.15	(27)
Auction markets	+1.39	(34)	-1.40	(1)	-2.63	(2)
Dealers	-4.55	(9)	-		-	
Mixed outlets	+6.65	(9)	-.59	(7)	-1.83	(2)
Total number sold	5,745	(100)	78,377	(100)	45,953	(100)

	Porkers		Cutters		Baconers	
Net price per kg dwt related to						
average of all herds (+ or -)	p	%	p	%	p	%
Direct sales	-.6	(49)	-2.7	(45)	+.2	(53)
Producers' groups	-1.6	(26)	+.4	(37)	0	(30)
Marketing companies	-1.4	(2)	-.2	(8)	-1.0	(5)
Auction markets[(c)]	+5.5	(10)	+7.0	(2)	-	
Mixed outlets	+2.3	(13)	+.3	(8)	0	(12)
Total number sold	65,985	(100)	135,308	(100)	81,022	(100)

(a) Excluding sales for breeding
(b) Prices adjusted to standard weights
(c) Prices converted to deadweight

Breeding stock sales and purchases

The cost of new or replacement breeding stock and their cull values when finished breeding are taken into account when calculating stock depreciation. For many years cull sow values have been influenced by exports to Germany where they are used for manufacturing pigmeat products. In times of high prices the value of these sows almost covered the cost of the replacement gilts, leaving minimal depreciation. Most cull breeding stock are sold by weight and the heavier animals usually make the highest prices. Feeding just to increase the weight of cull sows is usually uneconomic, however, unless very cheap feed is available, and is not recommended. Likewise, overfeeding sows during their lifetime is likely to be costly and could impair performance.

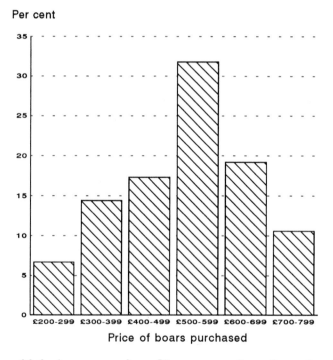

Figure 11.1 Average price of boars purchased per herd 1991 (percentage distribution)

The average cost of purchased boars in 1991 was £532. Herd averages ranged from £200 to just under £800 but individual boars cost up to £1,100. The range of boar prices for percentage distribution of herd averages is given in Figure 11.1. For standardisation of all herds prices included any haulage and insurance charges while in transit.

Some 32 per cent of herds buying boars returned average costs of between £500 and £600 per boar, 19 per cent were between £600 and £700.

Cull boars sold averaged £108 but ranged from £33 to £170. Prices were net of haulage and marketing charges.

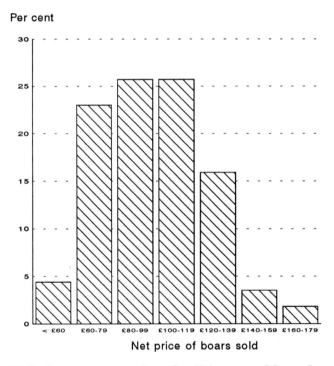

Figure 11.2 Average net price of cull boars sold per herd 1991 (percentage distribution)

Just over 90 per cent of the herds selling boars averaged between £60 and £140 per boar. The low prices often related to casualties or boars in poor condition, while a few of the high priced boars were intended for further breeding.

Herd average prices for gilts purchased were more uniform than boars with nearly 80 per cent between £140 and £180 each. The 5 per cent averaging under £120 were mostly gilts of an age younger than normal. A few herds bought only grandparent stock to breed their own replacements and these were the most expensive.

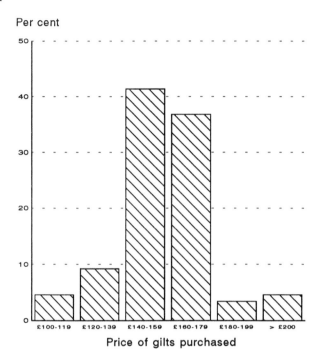

Per cent

Price of gilts purchased

Figure 11.3 Average price of gilts purchased per herd 1991 (percentage distribution)

The last of the breeding stock transactions deals with the disposal of cull sows. Nearly half of the sows sold fetched between £100 and £120. Prices reflected weights and condition of sows. They ranged from £65 for smaller or casualty sows to over £140 for heavier sows in good condition. While prices for purchased gilts reflects breeding performance potential; prices for culls reflect meat value only.

Prices received for cull breeding stock in 1991 fell from the record levels of the previous year, while the cost of new stock continued to rise. Consequently, depreciation rates were higher for the year, especially for herds buying in their replacements.

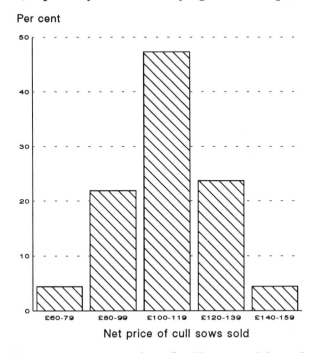

Figure 11.4 Average net price of cull sows sold per herd 1991 (percentage distribution)

Replacement gilts

Over the years the survey has sought to keep pace with technical developments by undertaking supplementary studies, comparing the performance achieved from various aspects of pig production. These were periodic investigations of certain systems and practices being performed on the farms in the scheme, which were already providing detailed information about their herds.

In the early days of the scheme, pig producers either bred their own replacement gilts, or bought them in, usually from a local breeder. There were many independent breeders, most quite small by present day standards. Normally purchased gilts would be guaranteed in-pig and due to farrow within a month. Quality varied and most breeders acquired their reputations by displaying their stock at agricultural shows and through word of mouth.

Table 11.3 A comparison of average breeding herd results for purchased gilts and own-bred gilts 1991

Breeding	Purchased	Own-bred[a]
Number of herds	63	30
Number of sows per herd[b]	175	148
Litters per sow in herd	2.29	2.22
Age at weaning (days)	25	27
Live pigs born per litter	10.6	10.6
Weaners per litter[c]	9.1	9.0
Weaners per sow[c]	21.0	20.0
Weaners per sow 1990	(20.9)	(20.5)
Weaners per sow 1989	(21.8)	(20.6)
Weaners per sow 1988	(22.1)	(20.4)
Weight of weaners[c]	18.2 kg	17.9 kg
Sow feed per weaner	60.8 kg	65.6 kg
Piglet feed per tonne	17.0 kg	17.5 kg
Cost of sow feed per tonne	£147.58	£145.36
Cost of piglet feed per tonne[c]	£264.16	£262.09
Compounds as % of total feed	73 %	65 %
Cost per weaner to 8 wks	£	£
Feed	13.42	14.13
Labour	5.61	5.99
Other costs	5.89	5.74
Stock depreciation	1.45	.15
Total breeding costs	26.37	26.01
Total breeding costs 1990	(£25.07)	(£24.19)
Total breeding costs 1989	(£23.62)	(£22.95)
Total breeding costs 1988	(£22.55)	(£22.42)

(a) Some grandparent stock purchased occasionally
(b) Monthly average (including in-pig gilts)
(c) At eight weeks of age

Later, as herd sizes increased and demand for gilts grew, breeding companies were formed, often from the amalgamation of a few independent breeders. The successful companies soon expanded and were doing enough business to employ geneticists and specialist marketing personnel. Most established multiplying herds for large-scale production of gilts, which were usually sold unserved. By then most of the independent breeders had given up or switched to commercial production of pigs for slaughter. In 1991, some three-quarters of the gilts joining herds in the scheme were purchased, mainly from breeding companies. Of the total number of breeding herds, 53 per cent purchased gilts, 25 per cent reared their own and 22 per cent did both.

From 1988 to 1991 comparisons of results were made of herds buying in gilts with those breeding their own. The herds doing both were omitted. The breeding herd results are shown in Table 11.3.

Twice as many herds bought in replacement gilts as reared their own in 1991. On average, they were slightly larger herds, produced marginally more litters per sow in herd, and weaned earlier. There was little difference between the two groups in litter size, both at birth and at eight weeks, but the extra litters per sow in herd produced 21 weaners per sow for purchased gilts compared with 20 for own-bred. The purchase gilts group also produced more weaners per sow in each of the three previous years. Usually, these weaners were marginally heavier at eight weeks than those from own-bred gilts.

Total sow feed used was slightly less for purchased gilts (35 kg per sow a year) and when shared by the number of weaners produced, the average per weaner was only 60.8 kg compared with 65.6 kg for own-bred gilts. In each of the three previous years less sow feed was used per weaner produced from the purchased gilts. They also consumed less piglet feed in 1991 and 1988 but in the two other years, the quantity per weaner was virtually the same. The cost per tonne of both types of feed was marginally less for own-bred gilts in 1991 and in two of the three previous years, probably reflecting that slightly more was home-mixed.

Feed costs per weaner were on average lower for the purchased gilts throughout, due mainly to higher number of weaners produced per sow in herd. Labour and other costs per weaner were similar for both groups in each of the four years. The main difference, of course, was in the stock depreciation

Table 11.4 A comparison of feeding herd results between those using
purchased gilts and own-bred gilts 1991

Feeding stock[a]	Purchased	Own-bred
Liveweight of pigs produced	81 kg	82 kg
Daily liveweight gain	.63 kg	.63 kg
Mortality percentage	2.7 %	2.5 %
Feed conversion rate[b]	2.58	2.63
Feed conversion rate 1990	(2.52)	(2.63)
Feed conversion rate 1989	(2.60)	(2.66)
Feed conversion rate 1988	(2.65)	(2.76)
Cost of meal per tonne	£156.01	£151.67
Cost of feed per tonne[b]	£153.47	£149.97
Compounds as % of total feed	40 %	33 %
Costs per kg liveweight gain	p	p
Feed	39.6	39.5
Labour	5.0	5.1
Other costs	5.6	5.8
Mortality charge	<u>1.2</u>	<u>1.0</u>
Total feeding costs	<u>51.4</u>	<u>51.4</u>
Total feeding costs 1990	51.3 p	51.9 p
Total feeding costs 1989	50.8 p	51.4 p
Total feeding costs 1988	47.7 p	49.4 p

(a) Excludes mainly breeding herds selling weaners
(b) Includes other feed (mainly by-products) "converted" to meal equivalent

charge, due to the cost of purchased gilts being higher than the
value of own-bred gilts entering the breeding herd. The cost of
purchased gilts would include the breeder's profit whereas the
value of own-bred gilts does not. Own-bred gilts were usually
transferred to the breeding herd at about 90 kg liveweight and,
for the purpose of calculating stock depreciation, were valued at
bacon pig price plus a nominal sum to allow for additional costs
involved. The price received for cull sows sold would frequently
exceed this transfer value, leaving boar transactions and sow

deaths to form most of the herd depreciation. In 1991, the depreciation charge per weaner was £1.45 for purchased gilts and 15p for own-bred gilts, a difference of £1.30 per weaner. This compared with differences of £1.28, £1.02 and 72p for the three previous years.

The higher stock depreciation charge for purchased gilts more than offset the savings achieved under other headings. As a result, their total costs per weaner were slightly higher but the weaners fractionally heavier. After allowing for weight variation, the difference in costs of producing weaners between the two groups was quite small. The purchased gilts did, however, produce more weaners and if these were valued, the total margin per sow would be higher for this group.

Both groups reared the progeny to finish at about the same weight. Details are given in Table 11.4.

Mortality was higher for purchased gilts in 1991 but in two of the three previous years, the position was reversed with own-bred gilts the higher. The key factor of feed conversion rate was best for purchased gilts throughout but as the feed was more expensive, for no apparent reason, there was little difference between the two groups in cost of feed per kg liveweight gain. In total the labour, other costs and mortality charges were also similar. Total costs were the same at 51.4p per kg for both groups in 1991 but were lower for purchased gilts in each of the three previous years, due mainly to better feed conversion rates.

The overall costs of breeding and feeding for own-bred gilts were just 36p per pig less than purchased gilts in 1991. A four year average reduced the difference to an insignificant 4p per pig. The advantage gained by purchased gilts, from more pigs per sow in herd and lower feed requirements, was offset by a high stock depreciation charge and the feed was more expensive per tonne. Both groups ended up with virtually the same production costs and, as the price received for finished pigs (Table 11.5) was also the same, there was no difference in profitability.

If the group purchasing gilts could have achieved the same performance by using feed at the lower price paid by the group using own-bred gilts, costs would have reduced by about 80p per pig. This saving would have reversed the comparative position of the two groups to make overall costs for purchased gilts on average 54p per pig less than own-bred gilts, still an insignificant amount.

**Table 11.5 A comparison of finished pig prices
for purchased gilts and own-bred gilts**

	Purchased	Own-bred
Net price per deadweight	p	p
1991	99.1	98.8
1990	117.5	118.0
1989	104.5	104.5
1988	87.9	87.7
Four year average	102.25	102.25

Outdoor versus indoor

Until the early 1980s, outdoor pig breeding herds had largely been out of favour in most parts of the country except the Southern counties for some 30 years. Outdoor production had otherwise been undertaken by only a small proportion of pig keepers. Herds were smaller and, to avoid the extremes of climate, sows were usually batch-farrowed in spring and autumn. Sales were, therefore, restricted to twice a year and the financial profits depended on prevailing prices at time of sale. Most arable farmers preferred to grow crops on their land rather than to keep pigs on it. Fields used for pigs were often occupied for several years and any disease problems were usually associated with "pig-sick" land.

The resurgence of outdoor production owed much to the development of hardy sows, capable of breeding outside all the year round, and to a national interest in the welfare of farm animals. Furthermore, this was a comparatively low cost system to establish. The number of pig producers was continuing to decline and many of the remaining herds kept expanding. With a plentiful supply of suitable farm transport, remoteness was no longer a handicap. In fact, environmental issues supported outdoor units, the more isolated the better. The main criteria were simply free draining land and a supply of drinking water. Most fields used for pigs are now changed every few years.

Sufficient outdoor units had been recruited into the scheme by 1984 to allow a study to be undertaken. The latest results for 1991 together with averages for the five years 1987-91 are shown in the following tables.

Table 11.6 A comparison of breeding results for outdoor and indoor herds to eight weeks

| | Outdoor | | Indoor | |
	1991	1987-91	1991	1987-91
Number of herds	11	10	108	108
Number of sows in herd[a]	243	223	155	144
Litters per sow in herd	2.15	2.15	2.28	2.28
Age at weaning (days)	28	30	26	27
Live pigs born per litter	10.4	10.5	10.7	10.7
Weaners per litter	8.9	9.1	9.1	9.2
Weaners per sow in herd	19.2	19.6	20.8	21.1
Weight of weaners[b]	18.7 kg	18.8 kg	18.0 kg	18.1 kg
Sow feed used per sow in herd	1.40 t	1.39 t	1.27 t	1.26 t
Sow feed used per weaner	72.9 kg	70.7 kg	61.2 kg	59.9 kg
Piglet feed used per weaner[b]	16.9 kg	17.6 kg	17.4 kg	17.7 kg
Cost of sow feed per tonne	£148.29	£143.45	£146.94	£142.58
Cost of piglet feed per tonne	£234.49	£219.74	£271.49	£257.58
Compounds as % of total feed	93 %	92 %	66 %	64 %

(a) Monthly average (including in-pig gilts)
(b) At eight weeks of age

This small sample of outdoor herds, when compared with other herds in the scheme, produced fewer pigs per sow and required more feed. They were larger herds, weaning two to three days later than indoor herds, and farrowed fewer litters. Fewer pigs were born and reared to eight weeks per litter in 1991, though piglet feed requirements were lower per pig. The five year average figures for these factors were, however, similar for both outdoor and indoor groups. The outdoor units had heavier pigs at eight weeks of age.

Sow feed consumption 10 per cent higher for outdoor herds and when shared by fewer weaners, was some 11 kg per pig more than the quantity required for the indoor herds. There was little difference in the cost of sow feed per tonne but piglet feed used by outdoor units was substantially cheaper. On balance, the outdoor units had a higher feed cost per weaner produced (Table 11.7) because they used more feed. If their piglet feed cost the same per tonne as the indoor herds, then feed costs per weaner would have been 63p more.

Table 11.7 A comparison of weaner costs for outdoor and indoor herds

| | Outdoor | | Indoor | |
	1991	1987-91	1991	1987-91
Cost per weaner[a]	£	£	£	£
Feed	14.61	13.82	13.70	13.09
Labour	6.17	5.12	5.76	5.02
Other costs				
Farm transport	.73	.59	.39	.34
Vet and vet supplies	.68	.44	.84	.74
AI fees	.08	.05	.11	.09
Power and water	.33	.31	1.25	1.07
Miscellaneous	.58	.43	.64	.53
Litter	.20	.15	.33	.27
Maintenance	.59	.36	.58	.48
Equipment charge	.29	.23	.29	.23
Buildings charge	.91	.66	1.59	1.49
Pasture charge	.64	.54	-	-
Total other costs	[5.03]	[3.76]	[6.02]	[5.24]
Stock depreciation	.85	.52	1.16	.90
Total costs[b]	26.66	23.22	26.64	24.25
Estimated value per weaner	27.35[c]	26.95[d]	27.00[c]	26.60[d]
Margin per weaner	.69	3.73	.36	2.35
Margin per sow a year	£13.25	£73.11	£7.49	£49.58

(a) To eight weeks of age
(b) No charge has been included for interest on capital
(c) Weaners valued at £27 for 18 kg, plus or minus 50p per kg
(d) Weaners valued at £26.55 for 18 kg, plus or minus 50p per kg

The outdoor units also had higher costs for labour and farm transport. Their costs were considerably less for buildings, veterinary, litter, power and water, which offset the higher feed costs and the land charge to give virtually the same total costs per weaner in 1991. The degree of confidence that can be placed on one year's results from a small sample is not great and a better appraisal is likely from the comparison shown for five years. Feed, labour and farm transport were still higher for the outdoor units while buildings, veterinary, litter, power and water were lower, but the different amounts involved for the five years 1987-91 resulted in total costs of £23.22, compared with £24.25 for the indoor herds. Total costs for both groups in 1991 were some £2 to £3 per weaner higher which reflects the rise in costs, especially for labour and other costs.

Stock depreciation charges were also lower for the outdoor units, both in 1991 and for the five years 1987-91. Prices received for cull boars and sows were virtually the same, though breeding stock mortality was marginally less. The main reason for smaller depreciation charges was lower costs of new and replacement boars and gilts. Boars for the outdoor units cost on average £52 less and gilts £14 less over the five years. This was probably due to advantages of scale by the larger outdoor units and buying in slightly younger stock. They also used more of their own-bred gilts, 34 per cent of the total brought in, compared with 23 per cent for indoor herds.

The outdoor units not only produced lower cost weaners but they were also heavier and, therefore, more valuable. The estimated value of weaners allowed 35p for the extra 0.7 kg in weight. With higher values and lower costs, the outdoor herds achieved the best margins. Even though they produced fewer weaners and used more feed, the margins per sow in herd a year were also much better.

The progeny of outdoor herds are usually sold as weaners and another comparison worth undertaking is to time of sale. Most weaners were sold at weights of around 30 kg and the following comparisons use actual prices received at sale. Table 11.8 shows the production results for herds selling weaners.

As with all breeding herds, the outdoor units producing weaners for sale were larger, farrowed fewer litters per sow a year and weaned two or three days later than indoor herds. Here, however, there was little difference in litter size and especially in the number of weaners per litter at sale. With fewer litters per sow the outdoor herds produced fewer pigs per sow in herd. At sale, they were usually slightly heavier, though not in 1991. Substantially more sow feed was used but piglet feed requirements at around 40 kg per weaner were about the same as for indoor herds (and in 1991 alone they were lower). Here too, the cost per tonne of piglet feed was lower for outdoor herds. Sow feed was also cheaper but by a smaller amount.

Table 11.8 A comparison of production results for outdoor and indoor herds selling weaners

| | Outdoor | | Indoor | |
	1991	1987-91	1991	1987-91
Number of herds	7	6	30	29
Number of sows in herd[a]	234	221	180	158
Litters per sow in herd	2.18	2.18	2.34	2.34
Age at weaning (days)	27	29	25	26
Live pigs born per litter	10.4	10.7	10.6	10.6
Weaners per litter	9.1	9.2	9.2	9.2
Weaners per sow in herd	19.6	20.1	21.3	21.5
Weight of weaners[b]	28.9 kg	30.0 kg	29.6 kg	28.4 kg
Sow feed used per sow in herd	1.43 t	1.40 t	1.26 t	1.25 t
Sow feed used per weaner	72.3 kg	69.7 kg	58.7 kg	57.8 kg
Piglet feed used per weaner	36.3 kg	40.3 kg	41.2 kg	39.8 kg
Cost per sow feed per tonne	£148.82	£141.79	£149.72	£145.17
Cost of piglet feed per tonne	£204.59	£192.81	£221.48	£214.85
Compounds as % of total feed	95 %	94 %	86 %	74 %

(a) Monthly average (including in-pig gilts)
(b) At sale

The costs of producing these heavier weaners to sale are shown in Table 11.9.

Table 11.9 A comparison of weaner costs to sale for outdoor and indoor herds

	Outdoor		Indoor	
	1991	1987-91	1991	1987-91
Costs per weaner[a]	£	£	£	£
Feed	18.11	17.45	18.08	17.08
Labour	6.91	5.65	6.40	5.55
Other costs				
Farm transport	.92	.80	.52	.45
Vet and vet supplies	.76	.47	.82	.74
AI fees	.10	.06	.07	.07
Power and water	.18	.16	1.20	1.07
Miscellaneous expenses	.65	.51	.86	.71
Litter	.22	.15	.37	.30
Maintenance	.56	.39	.63	.59
Equipment charge	.35	.30	.34	.28
Buildings charge	.90	.74	1.78	1.69
Pasture charge	.75	.68	-	-
Total other costs	[5.39]	[4.26]	[6.59]	[5.90]
Stock depreciation	1.22	.71	1.31	.98
Total costs[b]	31.63	28.07	32.38	29.51
Weaner price (net)	31.88	32.48	32.54	31.82
Margin per weaner	.25	4.41	.16	2.31
Margin per sow in herd a year	£4.89	£87.33	£3.41	£49.87
Return on capital	.9 %	18.8 %	.5 %	7.0 %
Capital requirements per sow	£	£	£	£
Value of sow	132	124	134	128
Share of boar's value	22	16	17	15
Buildings and equipment	256	215	463	429
Working capital	127	117	132	122
Total capital[c]	537	472	746	694

(a) To sale
(b) No charge has been included for interest on capital
(c) Excludes capital value of land

Costs cover the period to time of sale and as they were heavier pigs than those at eight weeks of age in Table 11.6, costs per weaner were higher. For outdoor herds producing weaners for sale, total costs in 1991 averaged £31.63 for weaners weighing 28.9 kg. The low prices received in 1991 only just covered costs, leaving a meagre margin of 25p per pig, whereas, for indoor herds, costs were on average 75p higher for slightly heavier weaners of 29.6 kg. Receipts were also higher but here again only just exceeded costs to give a margin of 16p per pig.

The weight of weaners sold has varied by a kilo or two over the years and the more realistic figures for the five years 1987-91 show that outdoor weaners were on average marginally heavier at 30 kg, compared to 28.4 kg from indoor herds, and therefore more valuable. Despite their greater weight, the outdoor weaners cost less to produce (£28.07 each) than the lighter indoor weaners (£29.51 each). The better prices received on average over the five year period left margins of £4.41 per pig for outdoor weaners, nearly double the £2.31 for indoor weaners.

Over the five year period, the outdoor units produced fewer pigs per sow and required more, but less expensive, sow feed. Piglet feed was also cheaper. Their combined feed costs per weaner were similar to the indoor herds, after allowing for the heavier weight of weaners. Only the pasture and farm transport charges were higher and these were more than offset by lower charges for buildings and power and water. Several items of other costs were marginally lower. Together, these savings gave them lower costs per weaner.

Although the outdoor units produced fewer pigs, the margin per sow a year was considerably higher. As expected, they had lower capital requirements which resulted in a substantially better return on capital. Low capital investment for housing and equipment is a major advantage and profitability compares favourably with indoor herds. Productivity often tends to be low and herd management is not easy, especially in the larger units. Barren sows and sterile boars may remain in the herd for some time before they are recognised. The success of the herd depends largely on the skill and ability of those looking after the pigs. A good man often overcomes the problems and difficulties associated with wet winters but a poor one will fail in ideal conditions. The availability of suitable land is, of course, the key factor for outdoor units but in regard to these comparisons a couple of other points should be considered. First, the sample was very small. Second, the winter weather was kind for most of this period and similar performance may be difficult to achieve in a severe winter.

Indoor dry sow accommodation

The introduction of electronic sow feeders and the proposed ban on sow stalls and tether systems led to a small study being undertaken in 1991 of dry sow accommodation. Lack of capital for investment and the uncertainty of profitable pig production have restricted the installation of electronic sow feeders but numbers seem to be steadily increasing. Sow stalls have existed for several years and, as herds expanded, this method was recommended for ease of management and control of the unit. By the 1980s objections on animal welfare grounds were being raised.

Table 11.10 Breeding results related to type of indoor dry sow accommodation 1991

	Electronic sow feeders	Stalls & tethers	Other[c] indoor
Number of herds	8	27	66
Number of sows in herd[a]	119	155	133
Litters per sow in herd	2.24	2.29	2.28
Age at weaning (days)	24	25	27
Live pigs born per litter	10.6	10.7	10.7
Weaners per litter	9.2	9.2	9.1
Weaners per sow in herd	20.5	20.9	20.8
Weight of weaners[b]	19.4 kg	18.1 kg	18.1 kg
Sow feed used per sow in herd	1.25 t	1.26 t	1.28 t
Sow feed used per weaner	61.2 kg	60.2 kg	61.2 kg
Piglet feed used per weaner[b]	16.8 kg	17.7 kg	17.6 kg
Cost of sow feed per tonne	£144.88	£149.91	£144.69
Cost of piglet feed per tonne	£278.05	£274.89	£263.00
Compounds as % of total feed	69 %	67 %	63 %

(a) Monthly average (including in-pig gilts)
(b) At eight weeks of age
(c) Includes yards, kennels, cubicles, pens and runs, etc

The results of 8 herds in the scheme using electronic sow feeders and of 27 using stalls were compared with 66 herds in other accommodation, such as yards, kennels, cubicles, pen and run, etc. Herds using more than one of the three systems were omitted, together with, of course, the outdoor herds.

The production results in Table 11.10 were remarkably similar for the three groups. Weaners from the herds with electronic sow feeders were heavier at eight weeks of age, while the group using stalls had more expensive sow feed. Otherwise, there was little difference between them in the number of litters per sow, litter size, weaners per sow in herd and feed requirements. Costs per weaner are given in Table 11.11.

Table 11.11 Costs per weaner related to type of indoor dry sow accommodation 1991

	Electronic sow feeders	Stalls & tethers	Other[d] indoor
Costs per weaner[a]	£	£	£
Feed	13.53	13.90	13.49
Labour	5.47	5.50	5.80
Other costs	5.81	6.31	5.78
Stock depreciation	1.19	1.25	1.11
Total costs[b]	26.00	26.96	26.18
Estimated value per weaner[c]	27.72	27.02	27.07
Margin per weaner	1.72	.06	.89
Margin per sow in herd a year	£35.26	£1.25	£18.51
Return on capital	5.1 %	.2 %	2.9 %
Capital requirements per sow	£	£	£
Value per sow	134	132	133
Share of boar's value	17	17	18
Buildings and equipment	442	382	375
Working capital	104	108	106
Total capital	697	639	632

(a) To eight weeks of age
(b) No charge has been included for interest on capital
(c) Weaners valued at £27 for 18 kg, plus or minus 50p kg
(d) Includes yards, kennels, cubicles, pens and runs, etc

Costs per weaner to eight weeks of age were marginally higher for the group using sow stalls or stalls and tethers, due mainly to more expensive feed and higher other costs. The value of weaners barely covered costs to give small margins for both per weaner and per sow a year. The group using electronic sow feeders had the lowest costs per weaner and, as they were heavier, the most valuable weaners to provide the best margins. Capital requirements per sow were higher due to buildings and equipment values but with the best margin per sow, they achieved the best return on capital.

These results are not greatly different but may indicate to those pig producers forced to abandon their sow stalls that profitability from other systems of sow accommodation should be just as good. The major difficulty for most will be funding the change. Many will have to scrap equipment capable of several years further service if they wish to stay in pigs. Financial aid seems unlikely, despite the encouragement that many producers received to install sow stalls. Perhaps it is not too late to reconsider and provide some assistance towards the cost of a change brought about in the name of animal welfare.

Chapter 12

Production Efficiency

Improvements in efficiency

The exodus of farmers from pigs which reduced the number of units in England and Wales from nearly 150,000 in 1955 to under 13,300 by 1992 is likely to have contributed to the considerable improvement in production efficiency over the years. In many cases the disbanded herds were the least successful and their departure helped raise averages. Those that remain now produce pigs more efficiently than before. The key factors of the number of pigs produced per sow in herd and the amount of feed required, in both breeding and finishing stages of production, have shown remarkable improvement.

Compared with 1950, feed requirements in 1990 for breeding and feeding per kg deadweight have improved by 40 per cent (25 per cent since 1970). Sows were producing on average 60 per cent more pigs per year (22 per cent since 1970) and this higher number provides a wider division of the sow's feed to give a lower share per weaner. Improved environment, health of the pigs and quality of feed have all made substantial contributions to increased efficiency, but the most important influence on performance comes from stockmanship, the ability and skill of those attending to the pigs to maximise achievement under prevailing conditions.

Improvement in feed requirements has been a notable achievement and helped to contain production costs when prices were rising fast. After many years of relative stability, the escalation in feed prices from an average of £35 per tonne in 1972 to £168 in 1984, an increase of 380 per cent in 12 years, would otherwise have been catastrophic for pig production as the increase in pig prices was much less.

152

Table 12.1 Average results

	1950	1960	1970	1980	1990
Breeding					
Litters per sow in herd	1.66	1.74	1.95	2.14	2.26
Live pigs born per litter	9.3	9.8	10.3	10.3	10.6
Weaners per litter	7.7	7.9	8.6	8.8	9.1
Weaners per sow	12.8	13.7	16.8	18.9	20.6
Feed used per weaner[a]- kg	119	116	95	82	79
Feeding					
Livewt of pigs produced - kg	90	85	87	81	82
Mortality percentage	3.4	3.1	3.0	2.5	2.4
Feed conversion rate[b]	4.7	4.0	3.7	3.2	2.6
Breeding and feeding					
Total feed used per kg deadweight (kg)	6.9	6.1	5.5	4.8	4.1

(a) To eight weeks
(b) From eight weeks

Between 1970 and 1991, meal prices in current terms increased by just over 400 per cent but largely as a result of improvements in efficiency, production costs increased by 309 per cent while finished pig prices increased by 265 per cent. Fluctuating pig prices in recent years have made single year comparisons unsound, as demonstrated by the difference in price per kg deadweight between 1990 and 1991. For this reason an average of three years provides a better guide to the changes in finished pig prices. Compared with the 1970-72 average of 28.5p per kg, prices in 1989-91 had risen to 106.7p, an increase of 274 per cent, which was slightly less than the increase in production costs.

In real terms, with the earlier years reflated to terms of money of 1991 purchasing power, prices as shown in Table 12.2 have fallen in most years for feed and total costs since 1977. Pig prices have also declined overall for a slightly longer period, but have been more variable due to the fluctuations. Indices of prices in real terms, related to the three years of 1970-72 to even

Table 12.2 Average costs and prices in current and real terms with indices

	Year[a]	Total meal[b] price per tonne	Total costs per kg deadweight	Finished pigs net price per kg dwt
		£	p	p
Current values	1970	32.07	23.9	27.0
	1975	70.54	47.9	55.7
	1980	126.71	79.9	84.2
	1985	155.67	90.0	101.0
	1990	160.85	93.7	117.3
	1991	161.50	97.7	98.7
Real terms[c]	1970	232.87	173.5	196.1
	1975	288.62	196.0	227.9
	1980	259.63	163.7	172.5
	1985	220.51	127.5	143.1
	1990	172.58	100.5	125.9
	1991	161.50	97.7	98.7
Indices of real terms 1970-72 = 100	1970	98.4	102.2	102.8
	1975	121.9	115.6	119.5
	1980	109.7	96.5	90.4
	1985	93.2	75.2	75.0
	1990	72.9	59.3	66.0
	1991	68.2	57.6	51.7
Average 1989-91		73.6	60.4	60.5

(a) Year ended 30 September
(b) Meal only (excludes other feed)
(c) Reflated by the Retail Prices Index at 1991 values

out the fluctuations, show that for the three years 1989-91 meal prices were 73.6 per cent of those in the base period, while pig prices had fallen to 60.5 per cent of that level. Fortunately, the improvements achieved in performance standards meant that on average the index of production costs moved similarly to that of pig prices. On this evidence, the costs and returns for the producers that remained in pigs have fallen roughly in line with each other. Without the improvements in efficiency, production costs would have been higher and for much of the time in recent years, pigs would have been unprofitable for most producers.

Seasonal variation

Standards of performance are usually better in the summer than during the winter. More weaners are produced per sow in herd, mortality is lower and feed requirements are less in both the breeding and feeding stages of production. The timing of the Cambridge scheme, where the winter half year ended on 31 March and the summer half year on 30 September, allowed comparisons to be undertaken between the two periods. Table 12.3 shows the superiority of the summer performance over the winter in five year averages.

Table 12.3 Variation in average results between winter and summer (summer minus winter)

Five year average	More weaners per sow	Less feed per weaner	Lower mortality	Better feed conversion rates
	No	kg	%	kg
1970-74	.50	4.8	.28	.14
1975-79	.42	3.8	.16	.17
1980-84	.46	4.6	.18	.15
1985-89	.20	3.2	.16	.07
1990 & 91[a]	.20	2.0	.05	.04

(a) Two years

As results have improved over the years, the difference between winter and summer performance has narrowed. Until 1985 nearly half a pig more was produced per sow in herd during the summer than the winter. Requirements of feed for breeding stock were some four kilograms less per weaner. Mortality in the feeding stage was about 0.2 percentage points less (approximately 2.4 per cent compared with 2.6 per cent) and feed conversion rates about .15 kg of feed per kg liveweight gain better - or 4 per cent (3.35 instead of 3.50). Over this period the savings in feed between winter and summer on average totalled some 15 kg in producing a 90 kg liveweight pig, which at £150 was worth £2.25. Further savings from labour and other costs during the summer made a total of about £3 per pig.

More recently, climatic changes have reduced the difference between winter and summer performance. The mainly mild winters and hot summers have cut the summer advantage by about a half, to around £1 to £1.50 per pig. If the weather continues to provide kind winters to the advantage of pig production, especially from the increasing number of outdoor breeding herds, costs in the summer are likely to remain about £1 to £1.50 per pig (or 2 per cent) lower than in the winter. Should more severe winters return, the value of the difference between winter and summer results could again on average amount to some £3 per pig, or 4 per cent.

The quality of housing influences the feed requirements especially in winter. Pigs grow better in well insulated buildings with warm and dry sleeping areas, which are free of draughts. Good accommodation need not be expensive and a plentiful supply of clean straw can help to improve insulation. There can be little doubt that efforts to provide the pigs, especially feeding stock, with suitable conditions can often be justified to minimise the difference in performance and costs between winter and summer periods.

Earlier weaning

The increase in the number of litters per sow in herd a year reflects the move to earlier weaning. A few herds in the Cambridge scheme started weaning early in around 1965. Until then the traditional age for weaning had been eight weeks and on average breeding herds achieved 1.75 litters per sow of about eight weaners to produce around 14 pigs a year. By 1990, production had improved to 2.26 litters of nine weaners to give just over 20 pigs per sow a year. About one-third of the increase in the number of pigs produced per sow a year comes from the improvement in litter size and two-thirds from the additional litters per sow. Reduced mortality accounts for part of the improvement in litter size but the extra litters per sow come almost entirely from earlier weaning.

In 1970, only 10 per cent of herds in the scheme weaned at younger than five weeks of age; by 1990 this had risen to 83 per cent, with 8 per cent of them weaning at under three weeks (Table 12.4).

Table 12.4 Age at weaning - percentage of herds

	Under 3 weeks	3 & 4 weeks	5 & 6 weeks	7 & 8 weeks
1970	-	10	46	44
1974	-	15	67	18
1978	9	35	48	8
1982	14	55	29	2
1986	13	60	25	2
1990	8	75	14	3

Weaning at under three weeks was never very popular and has been losing favour in recent years. To start with, herds in this group in the scheme aimed to wean at two weeks but soon most had delayed until at least two and a half weeks. Many were dissatisfied with results and some voiced difficulties afterwards in managing to get sows successfully served again. The few remaining in 1990 were usually weaning at only a day or two under three weeks.

This was not the first attempt at weaning early. In the mid 1950s a national compounder developed a substitute feed for sow's milk and persuaded several breeders to try their system of weaning at ten days. Results were mixed and within a few years, all had returned to the then traditional weaning at eight weeks of age.

During the late 1960s weaning at five and six weeks became popular and this was followed by a few weaning at three weeks, which was made possible by the introduction of suitable feeds for the young pigs. After a few years most producers weaned their pigs at either three or five weeks but more recently several have changed to an average of four weeks. Nowadays there seems to be no definite recommended age and individual producers choose whatever age suits their circumstances. Much depends on the skill of the person attending to the young pigs and on their ability to make the system work satisfactorily. Whatever the age at weaning, it is vitally important to ensure that the sows conceive again shortly after weaning. It is wasteful to use expensive early weaning feeds if the herd fails to produce extra litters in the time saved.

A comparison of the 1991 breeding results by age at weaning is given in Table 12.5.

Table 12.5 Breeding results related to age at weaning 1991

Age at weaning (days)	18-24	25-31	32-38	39+
Number of herds	56	44	11	8
Number of sows in herd	202	143	130	46
Average age at weaning (days)	22	27	35	44
Litters per sow in herd	2.30	2.25	2.15	1.83
Live pigs born per litter	10.5	10.8	10.7	11.1
Weaners per litter at 8 wks	9.1	9.0	9.2	9.5
Weaners per sow in herd	20.9	20.3	19.8	17.3
Weight of weaners at 8 wks	18.0 kg	18.2 kg	19.0 kg	17.1 kg
Sow feed used per sow in herd	1.26 t	1.31 t	1.41 t	1.28 t
Feed used per weaner to 8 wks	kg	kg	kg	kg
Sow feed	60.3	64.8	70.8	74.2
Piglet feed	17.3	17.7	16.0	15.5
Total feed	77.6	82.5	86.8	89.7
Cost of sow feed per tonne	£147	£147	£143	£145
Cost of piglet feed per tonne	£272	£265	£230	£254

The results were as expected, apart from the quantity of piglet feed per weaner to eight weeks of age. Normally the amount used for 18-24 days weaning was slightly more than for 25-31 days. Usually the amount per weaner declined as the age of weaning increased. However, perhaps it would be surprising if results in a comparison of commercial enterprises always came out as expected.

Herd size was on average largest for the earliest weaning group, while the herds weaning at 39 days and over were comparatively small. The number of litters per sow in herd was greatest with the 18-24 days group and they produced more weaners per sow in herd, despite litter sizes, both born and reared, being slightly smaller than some others. Less sow feed was used in the early weaner herds and when this was shared by the number of pigs produced, the increase in quantity per weaner with age at weaning was even more noticeable. This was partly offset by more of the comparatively expensive piglet feed used with early weaning. Most of the outdoor units, where sow feed requirements were higher, were included in the 32-38 days weaning group. The cost of sow feed per tonne varied only marginally.

Feed, labour and total costs per weaner, shown in Table 12.6, tended to increase with age at weaning, though other costs were higher for the early weaning groups. Stock depreciation was highest for the 18-24 days group.

Table 12.6 Breeding costs related to age at weaning 1991

Age at weaning (days)	18-24	25-31	32-38	39+
Costs per weaner at 8 weeks	£	£	£	£
Feed	13.58	14.23	13.85	14.68
Labour	5.66	5.71	7.09	7.71
Other costs	5.93	6.00	5.33	4.54
Stock depreciation	1.26	.92	.97	.68
Total costs	26.43	26.86	27.24	27.61
Estimated value per weaner[a]	26.96	27.08	27.51	26.54
Margin per weaner	.53	.22	.27	-1.07
Margin per sow in herd a year	£11.08	£4.47	£5.35	-£18.51

(a) Weaners valued at £27 for 18 kg plus or minus 50p per kg.

Average weight of weaners at eight weeks of age varied marginally between the groups, which accounted for much of the difference in total costs per weaner between the three earliest weaning groups. Although they are not usually sold at this time, placing a value on them allows margins per weaner and per sow to be calculated. It should be remembered that 1991 was a year of low profitability due to poor pig prices, but as in most years margins per weaner for the earlier weaning groups differed by only a few pence. The 39 days and over group usually had the lowest margins. On a per sow basis, however, the earliest weaning gave the best margins because more weaners were produced per sow. This apparent advantage is, however, largely offset when capital requirements are taken into account. The assessed amount given in Table 12.7 includes the current value of buildings, the breeding stock and working capital for feed, labour and other costs.

Table 12.7 Capital requirements per sow 1991

Age at weaning (days)	18-24	25-31	32-38	39+
	£	£	£	£
Value of sow	134	132	130	123
Share of boar's value	17	19	16	14
Buildings and equipment	382	363	254	233
Working capital	106	108	113	120
Total capital	639	622	513	490
Return on capital	1.7 %	0.7 %	1.0 %	-3.8 %

Total capital estimates were highest for the earlier weaning herds mainly because of the higher investment in buildings and equipment. The low return on capital for all groups reflected the poor pig prices in 1991 and subsequent margins. If interest were paid on the total capital, the charge per weaner, or per sow, would exceed the margins shown for average performance.

A one year comparison of this type may be influenced by some chance results. Hence a further comparison of the key factors involved was undertaken for the previous five years (Table 12.8).

The average physical results by age at weaning for the five years 1986-90 are similar to those for 1991, apart from the smallest group weaning at 39 days and over. Few herds wean as late as this nowadays and the method is rather old-fashioned, so more attention should be given to the other groups. Slightly more feed was used per weaner in 1991 in each of the three groups but the relationship between them remained virtually the same. Costs per weaner were higher in 1991, due mainly to increases in the price of piglet feed per tonne and rising labour and other costs. When valued according to weight, the resulting margins for the five years between these groups again only differed by a few pence.

Table 12.8 Five years average breeding results related to age at weaning 1986-90

Age at weaning (days)	18-24	25-31	32-38	39 +
Litters per sow	2.32	2.22	2.12	1.83
Weaners per sow in herd	21.3	20.3	19.9	18.2
Weight of weaners at 8 wks	18.1 kg	18.2 kg	18.6 kg	17.8 kg
Feed used per weaner to 8 wks	kg	kg	kg	kg
Sow feed	58.4	62.8	67.3	77.6
Piglet feed	18.1	18.0	17.5	15.2
Total feed	76.5	80.8	84.8	92.8
Cost per weaner to 8 wks	£	£	£	£
Feed	12.84	13.29	13.26	14.30
Total	23.16	23.39	23.82	24.21
Estimated value per weaner	26.00	26.05	26.23	25.85
Margin per weaner	2.84	2.66	2.41	1.64
Total capital per sow	£587	£550	£482	£468
Return on capital	10.3 %	9.8 %	9.9 %	6.4 %

With more weaners produced, the earliest weaning herds achieved the best margins. They also required more capital, however, so the return on capital was not very different between the three groups.

To be successful with early weaning, it is essential that more weaners per sow are produced than with later weaning. Age at weaning alone will not ensure a high number of weaners per sow a year. Weaning earlier does no more than provide an opportunity to produce more. Indeed some herds weaning at five weeks achieve better results than others weaning at three weeks. It is the number of litters actually produced per sow a year that is really the important factor. This all starts, of course, with ensuring that sows are served properly at the right time.

Feed types

As feed forms by far the major part of costs in pig production, its efficient use is directly related to the profitability of the herd. Substantial improvements have already been made in feed conversion rates. Further improvements seem progressively harder to achieve, although considerable variation still exists between herds and possibilities exist for many to do better still. Feed prices have fallen less in real terms than pig prices, and improved efficiency has been necessary to maintain margins. Most likely this situation will continue in future, so producers must always strive to reduce production costs simply to keep pace with the long-term decline in pig prices.

Pig producers have little control over the prices of feed ingredients but there are marked differences between herds in the cost per tonne of total ration. Basically the cost depends on quality and on whether purchased compounds are used or feed is mixed on the farm. If by-products are available, this is another consideration in reducing the price per tonne for feeding stock. Price per tonne alone is not the crucial factor. It is the combination of price and quantity required to increase weight that is important. While many opt for low cost rations, others prefer to use better quality and more expensive feed for enhanced performance. The aim should be a low cost of feed per kg liveweight, no matter whether this is achieved from low requirements of expensive feed, or higher usage of comparatively cheap feed.

The scheme provided the opportunity of comparing results of herds using purchased compounds with those using feeds mixed on the farm. Charges for milling and mixing were included in the price of home-mixed feeds. With home-mixing it is also of interest to distinguish between herds using concentrate mixes and those using all straights. Herds using both compounds and home-mixed feeds were left out of these comparisons unless they were also using by-products.

Table 12.9 A comparison of breeding results by type of feed used 1991[a]

Breeding stock	Purchased compounds	Home-mixed using concentrates	Home-mixed using straights
Number of herds	67	23	22
Number of sows in herd[b]	181	118	142
Litters per sow in herd	2.26	2.24	2.24
Age at weaning (days)	26	28	26
Weaners per sow in herd	20.6	20.4	20.0
Weight of weaners at 8 wks	18.2 kg	18.1 kg	18.2 kg
Sow feed used per sow in herd	1.32 t	1.22 t	1.22 t
Sow feed per weaner to 8 wks	63.9 kg	60.1 kg	61.2 kg
Piglet feed per weaner to 8 wks[c]	17.1 kg	17.8 kg	17.6 kg
Cost of sow meal per tonne[d]	£150.33	£142.92	£139.53
Cost of piglet meal per tonne	£264.29	£274.69	£266.34
Costs per weaner to 8 wks	£	£	£
Feed	14.08	13.48	13.23
Labour	5.77	6.06	5.36
Other costs	5.76	6.18	6.22
Stock depreciation	1.24	.93	1.11
Total breeding costs	26.85	26.63	25.92
Feed used	%	%	%
Compounds - sows	79	-	-
Compounds - piglets	21	13	15
Concentrate mixes	-	16	-
Wheat	-	33	36
Barley	-	26	17
Milling offals	-	9	12
Soya bean and fish meal	-	-	10
Other concentrates (incl additives)	-	3	10
	100	100	100

(a) Excludes herds using compounds and home-mixed
(b) Monthly average (including in-pig gilts)
(c) Creep feed was usually purchased
(d) Milling and mixing charges included

Type of feed appears to have little influence on performance in the breeding stage of production as shown in Table 12.9. The quantity of sow feed used by the compounds group seems high in relation to others but ten of the 67 herds using compounds were kept outdoors and sows on this system normally consume more feed than those kept indoors. Once allowance is made for this, sow feed requirements both per sow and per weaner are on average similar to herds using home-mixed feed.

The one major difference was in the cost of sow feed per tonne. Sow compounds cost £7.41 per tonne more than home-mixed with concentrates and £10.80 more than home-mixed with straights. Lower feed costs per tonne for home-mixing with straights were mainly responsible for this group having lowest cost per weaner, though low labour costs helped. It was fairly common for home mixes to buy in creep and starter feeds as compounds for their piglets. Of the total piglet feed used, 55 per cent was purchased compounds for home-mixing with concentrates and 65 per cent for home-mixing with straights; the rest was mixed on the farm.

Breeding results related to type of feed used in previous years have seldom shown any distinct advantages for one group. The price of feed per tonne was always highest for compounds, as expected, but the quantities involved were usually too small to give much higher costs per weaner than others.

By contrast, at the feeding stage there were on average clear cost benefits for herds using home-mixed feed. Their cost of feed per tonne was less than that for purchased compounds, while the conversion rates were about the same for both the concentrates and the straights groups, to give them a lower feed cost per kg liveweight gain. The herds using by-products had an even lower cost of feed per tonne but this advantage was offset by a worse conversion rate (Table 12.10).

The results for the four groups were influenced by variation in the weight of pigs produced and brought in to the herd, which accounts for some of the difference in feed conversion rates. In 1991, the group using compounds had a worse than expected conversion rate, slightly inferior to both home-mixed groups. Usually compounds have achieved about 0.1 better conversion rate than the home-mixed groups. The main difference between groups was in the cost of feed per tonne, with compounds nearly £11 per tonne more expensive in 1991 than home-mixed with concentrates, £18 per tonne more than home-

Table 12.10 A comparison of feeding stock results by type of feed used 1991[a]

Feeding stock	Purchased compounds	Home-mixed with concentrates	Home-mixed with straights	Users of by-products
Number of herds	36	22	31	14
Pigs produced per herd	3027	1,913	3,306	5,072
Lwt of pigs produced	81 kg	82 kg	87 kg	91 kg
Lwt of pigs brought in	21 kg	20 kg	20 kg	25 kg
Daily lwt gain	.63 kg	.61 kg	.67 kg	65 kg
Mortality percentage	2.4 %	2.4 %	2.2 %	2.1 %
Feed conversion rate[c]	2.67	2.57	2.61	2.86
Cost of meal per tonne[b]	£163.35	£152.54	£145.45	£150.53
Cost of feed per tonne[c]	£163.35	£152.54	£145.45	£132.48
Costs per kg lwt gain	p	p	p	p
Feed	43.6	39.3	38.0	37.9
Labour	5.3	5.1	5.0	6.0
Other costs	6.3	5.4	5.8	7.2
Mortality charge	1.2	1.1	.9	1.1
Total feeding costs	56.4	50.9	49.7	52.2
Type of feed used	%	%	%	%
Compounds	100	-	-	17
Concentrate mixes	-	22	-	-
Wheat and barley	-	72	60	24
Milling offals	-	4	8	9
Soya bean & fish meal	-	-	22	17
Other concentrates[d]	-	2	10	9
Other feeds (by-products)	-	-	-	24
	100	100	100	100
1990 results				
Feed conversion rate	2.60	2.69	2.61	2.84
Cost of feed per tonne	£163.03	£152.25	£146.54	£128.42
Feed cost per kg lwt gain	42.4 p	40.9 p	38.2 p	36.4 p
Total cost per kg lwt gain	54.8 p	51.6 p	48.7 p	49.5 p

(a) Excludes herds using both compounds and home-mixed except users of by-products
(b) Milling and mixing charges included
(c) Includes other feed "converted" to meal equivalent
(d) Includes beans, cereal substitutes, molasses, fats and additives

mixed with straights and £31 per tonne more than those using by-products. The same order prevailed for feed costs per kg live-weight gain. Feed cost 43.6p per kg gain for compounds, some 4p more than home-mixed with concentrates and just over 5p more than home-mixed with straights and by-products. In 1990, with conversion rates better for compounds and worse for home-mixed with concentrates, the difference between these two in feed costs per kg gain was much less. The difference between compounds and home-mixed with straights was also reduced but by a smaller margin while the relationship with the by-products group was about the same. The variation in labour and other costs per kg was most likely a reflection of the circumstances on these farms rather than feed use. If the standard of buildings between the groups had affected feed conversion rates, then this would have been compensated for in the charges made for build-ings and included in other costs. Charges for home-milling and mixing have been added to the cost of feed but interest charges have not and these would raise costs more for home-mixing than for purchased compounds. An assessment of likely interest charges, based on a rate of 12.5 per cent on the current value of plant and working capital for one month, to cover feed stocks, labour and running costs, would add about £4 per tonne to feed costs for home-mixing and this would amount to about 1p per kg liveweight gain. While this interest charge would reduce the advantage for home-mixers, those using straights would still have the lowest total costs per kg liveweight gain both in 1990 and 1991.

The number of by-product users increased sufficiently to form a separate group in 1990 and 1991. By-products consisted mainly of cereal starch, skimmed milk and whey products plus small quantities of other waste feeds suitable for pigs. In nearly all cases continuity of supplies was a problem and changes in feeds used were frequent. The costs of equipment and storage tanks, often second-hand, were included under feed preparation.

Milling and mixing

Charges for milling and mixing feed on the farm were assessed according to individual circumstances. Equipment, buildings and storage were charged at 10 per cent of current value, irrespective of age. Units ranged from small mills and mixers, often old and low valued, housed in general purpose barns, to large, modern automated plants. The former usually had low depreciation charges but required more labour, while the newer specialised plants had their high depreciation charges partly offset by low labour requirements. A few prepared feed for other livestock in addition to pigs. The costs of repairs and replacement were included, together with the Pharmaceutical Society registration fees. Running costs included labour, power and an allowance for milling loss. In all cases the charges were made according to each farm's own costs and not to a standard figure.

Milling and mixing costs for all farms averaged £7.07 per tonne in 1991. Allowing for an interest charge, the total came to £10.86 per tonne. The 76 farms that prepared feed were divided into four groups according to tonnage milling and mixed per annum. As expected, costs per tonne declined as tonnage increased, ranging from £8.71 (excluding interest charges) for those milling and mixing less than 200 tonnes a year to £6.73 for over 800 tonnes. The larger units benefited from the extra tonnage over which to spread the depreciation. With more automation less labour was required and some of them bought electricity cheaper on the maximum demand tariff.

In most years, milling and mixing charges followed a similar pattern of costs per tonne as the larger units gained from the advantage of scale. Just occasionally, exceptional circumstances may have distorted this pattern, such as in 1988 when a few large herds in the over 800 tonnes group cut production because of low pig prices. Depreciation that year was shared by a lower than usual tonnage and raised costs per tonne. But such occasions are rare.

Many of the farms in the scheme have had their milling and mixing equipment for several years and values were correspondingly low. To replace or buy new at present day costs would be more expensive. At average efficiency, a 200 sow unit producing bacon weight pigs would require around 1,000 tonnes of feed per year, excluding piglet feed. With possible savings of £7 to £12 (after higher depreciation) and only minor differences in conversion rates, some compound users may be tempted to consider changing to home-milling and mixing. The savings on 1,000 tonnes a year should more than cover higher interest charges. To combat this possibility some compounders may offer additional discounts to make their prices more competitive with home-mixed, and to avoid losing a customer. Pig producers contemplating their own milling and mixing plant should also investigate any alternative for this investment. If efficiency is at a high standard, a similar investment in buildings and stock to expand the unit may be just as rewarding.

Table 12.11 Feed preparation costs per tonne 1991
(including feed milled and mixed for other livestock)

	Average of all farms	By size - tonnes per annum			
		<200	200-399	400-799	>800
	£	£	£	£	£
Depreciation	1.65	1.86	1.84	1.51	1.65
Repairs and fees	.58	.58	.64	.61	.54
Labour	2.51	3.52	2.55	2.60	2.35
Power	1.51	1.89	1.59	1.58	1.42
Milling loss	.82	.86	.87	.88	.77
Total (excl interest)	7.07	8.71	7.49	7.18	6.73
Interest charge[a]	3.79	4.03	4.01	3.54	3.70
Total (incl interest)	10.86	12.74	11.50	10.72	10.43
Total quantity (tonnes)	44,510	2,217	6,513	13,398	22,381
Tonnes per farm	586	139	310	558	1,492
Quantity used for pigs (tonnes)	40,817	2,177	6,072	12,471	20,097
Number of farms	76	16	21	24	15
Cost per tonne[b]	£	£	£	£	£
1986	5.36	7.66	6.53	5.95	4.80
1987	5.65	6.92	6.13	6.02	5.17
1988	6.40	7.61	6.16	6.01	6.63
1989	6.36	7.86	6.51	6.22	6.22
1990	6.63	8.47	6.89	6.82	6.29

(a) Interest charged at 12.5 per cent on the current value of buildings and equipment and working capital for one month to cover feed, labour and running costs
(b) Excluding interest charges

Current costs of milling and mixing have increased steadily since 1970 (see Table 12.12), but in real terms, at 1991 sterling values, they have fallen.

Table 12.12 Average milling and mixing costs per tonne

Year[a]	Depreciation	Repairs	Labour	Power	Milling loss	Total costs
Current values	£	£	£	£	£	£
1970	.34	.14	.43	.23	.25	1.39
1975	.38	.23	.73	.45	.48	2.27
1980	1.08	.42	1.29	.98	.63	4.40
1985	1.25	.58	1.77	1.34	.76	5.70
1990	1.57	.54	2.28	1.47	.77	6.63
1991	1.65	.58	2.51	1.51	.82	7.07
Real terms[b]	£	£	£	£	£	£
1970	2.47	1.02	3.12	1.67	1.81	10.09
1975	1.56	.94	2.99	1.84	1.96	9.29
1980	2.21	.86	2.65	2.01	1.29	9.02
1985	1.77	.82	2.51	1.90	1.07	8.07
1990	1.68	.58	2.44	1.58	.83	7.11
1991	1.65	.58	2.51	1.51	.82	7.07
Indices of real terms[c]						
1970	114	104	105	106	110	108
1975	72	96	101	116	119	100
1980	102	88	89	127	79	97
1985	82	84	85	120	65	86
1990	78	59	82	100	51	76
1991	76	59	85	96	50	76

(a) Year ended 30 September
(b) Reflated by the Retail Price Index at 1991 values
(c) 1970-72 = 100

Between 1970 and 1991 current costs of farm milling and mixing increased fivefold from £1.39 to £7.07 per tonne, the largest increases being in labour and power, though an increased tonnage per farm has meant more tonnes to share depreciation and repairs which help to contain costs. In real terms, the 1991 total costs per tonne have fallen on average to 76 per cent of the 1970-72 level.

Feed prices

Feed prices in current values (shown in Table 12.13) also increased by nearly fivefold between 1970 and 1991, except for protein meals, which were about three times higher. Compound prices in 1985 were remarkably similar to those of 1991, and this gives the appearance of stability. In fact, prices gradually fell and by 1988 were about £7 per tonne lower but they increased again in the following year. Similarly, wheat and barley prices were lower in 1988 but by 1991 were some 10 per cent above the 1985 level. Restricted supplies of soya bean meal in 1989 saw prices rise by £25 per tonne and then as imports returned to normal during the next two years, prices declined to even lower levels than previously. Fish meal prices, influenced by soya bean meal, also peaked in 1989 and by 1991 had returned to near the 1985 level.

Movements in compound prices appear justified when related to changes in cereal and protein meals. Compared to 1985, the dearer wheat and barley and cheaper soya bean meal in 1991 seem to have balanced reasonably well with changes in the price of compounds during this period. Advances in ration formulation and inclusion of additives have made feeder compounds far more expensive than sow compounds, whereas 20 years previously sow compounds usually cost more per tonne than feeder compounds.

In real terms, prices of the main feeds used in pig production have been falling almost continuously since 1977. By 1991, prices of compounds for sows were 59 per cent and for feeders 66 per cent of their levels in the 1970-72 base period. Wheat and barley prices were down to 68 and 67 per cent of base, while soya bean (38 per cent) and fish meal (48 per cent) had fallen even further. These differing rates of fall may suggest that price changes over this longer period have favoured home-mixing, as the greater fall in protein meal prices does not appear to be fully reflected in the more recent compound prices. Straight feeds are clearly identifiable but compounds are not and it will be argued that present day compounds are different from those of the early 1970s, so that direct comparisons are not possible. Compounds are likely to have improved considerably, yet production efficiency from home-mixed feeds has been maintained to retain their relative competitiveness.

Table 12.13 Average prices of purchased feed per tonne

Year[a]	Sow cmpds	Feeder cmpds	Wheat	Barley	Soya bean	Fish meal
Current values	£	£	£	£	£	£
1970	34.81	34.20	26.61	24.20	54.21	92.46
1975	73.39	74.54	55.53	56.64	90.33	136.54
1980	125.98	134.70	99.16	96.38	131.55	253.65
1985	148.60	164.39	113.42	111.37	159.83	301.98
1990	147.95	168.97	116.02	110.34	160.15	308.24
1991	149.95	167.74	124.97	120.43	145.15	297.57
Real terms[b]	£	£	£	£	£	£
1970	252.77	248.34	193.23	175.73	393.64	671.39
1975	300.28	304.99	227.21	231.75	369.59	558.67
1980	258.13	276.00	203.18	197.48	269.55	519.73
1985	210.49	232.86	160.66	157.76	226.40	427.75
1990	158.74	181.29	124.48	118.38	171.82	330.71
1991	149.95	167.74	124.97	120.43	145.15	297.57
Indices of real terms[c]						
1970	100	98	106	98	104	107
1975	119	120	124	129	98	89
1980	102	109	111	110	71	83
1985	83	92	88	88	60	68
1990	63	71	68	66	45	53
1991	59	66	68	67	38	48

(a) Year ended 30 September
(b) Reflated by the Retail Prices Index at 1991 values
(c) 1970-72 = 100

Chapter 13

Factors Affecting Profitability

The profitability of pig production varies considerably between herds, depending on the relationship between costs of production and the price received for pigs sold. The variation in costs is greater than in price received. Output is also important and this is measured by the number of pigs produced per sow for breeding herds and by number sold, or turnover, for feeding herds.

A comparison of performance in the following key factors will indicate how costs and returns are affected by standards achieved:

> Weaners produced per sow
> Breeding costs per weaner
> Feeding costs per kg liveweight gain
> Price per kg deadweight of sales

Weaners produced per sow

In 1991, just over 88 per cent of the herds in the scheme produced between 16 and 24 weaners per sow in herd a year; the remainder were spread fairly evenly below and above this range. Weaners per sow are dependent on the farrowing frequency and litter size, and the most successful herds do well in these factors. Herds were grouped according to the number of weaners produced (Table 13.1) and results clearly indicate the best herds achieved high standards throughout.

As the number of weaners produced per sow in herd a year increased, so did herd size, the number of litters per sow and litter size, both at birth and at eight weeks of age. The weight of weaners varied, roughly in line with the quantity of piglet feed consumed. The group producing 20 and 21 weaners per sow had the most expensive feed, 75 per cent of which was compounds. Although the quantity of sow feed used per sow varied, when shared by the number of weaners, the quantity per weaner produced declined quite dramatically as numbers increased, from 80.5 kg for the group with fewer than 18 weaners per sow to 55.3 kg for over 22 weaners. The difference of 25.2 kg at £145 per tonne was worth £3.65 per weaner and largely accounted for the difference in costs per weaner between the two groups.

Table 13.1 Breeding results related to the number of weaners produced per sow in herd 1991

Range - weaners per sow	<18	18 & 19	20 & 21	>22
Number of herds	17	33	44	25
Number of sows in herd[a]	95	159	178	184
Litters per sow in herd	2.00	2.19	2.29	2.38
Age at weaning (days)	32	26	25	25
Live pigs born per litter	9.9	10.4	10.6	11.1
Weaners per litter at 8 wks	8.0	8.7	9.1	9.7
Weaners per sow in herd	15.9	19.1	21.0	23.2
Weight of weaners at 8 wks	18.0 kg	18.4 kg	17.9 kg	18.2 kg
Sow feed used per sow in herd	1.28 t	1.26 t	1.31 t	1.28 t
Sow feed used per weaner	80.5 kg	66.1 kg	62.6 kg	55.3 kg
Piglet feed used per weaner	17.0 kg	17.8 kg	16.8 kg	17.7 kg
Cost of sow feed per tonne	£145.91	£145.60	£149.46	£145.25
Cost of piglet feed per tonne	£261.85	£263.59	£271.38	£264.43
Compounds as % of total feed	59 %	68 %	75 %	65 %
Costs per weaner at 8 weeks	£	£	£	£
Feed	16.19	14.32	13.86	12.72
Labour	6.05	6.36	5.73	5.37
Other costs	6.00	6.13	5.87	5.66
Stock depreciation	1.25	1.05	1.20	1.03
Total breeding costs	29.49	27.86	26.66	24.78
Margin per £100 output	-£7.26	-£0.50	£0.97	£8.67

(a) Monthly average (including in-pig gilts).

Costs per weaner at eight weeks also declined as the number of weaners per sow increased. For herds producing more than 22 weaners per sow a year, total costs averaged £24.78 compared with £29.49 for herds with fewer than 18 weaners per sow. Averages of the two intermediate groups at £26.66 and £27.86 give a good example of the effect that weaner numbers per sow have on production costs. The value of the weaners is likely to differ only by the marginal variation in weight. Herds producing the most weaners have the lowest costs and, therefore, are the most profitable.

Breeding costs per weaner

The findings of the previous comparison based on weaner numbers are confirmed in the following table (Table 13.2), relating to costs per weaner. The breeding herds were again selected into four groups, this time according to their average production costs. The lowest cost group produced weaners at less than £24 each and the highest had costs of over £30.

The group with the lowest costs per weaner produced the most weaners per sow in herd a year, from more frequent farrowings of larger litters than other groups. This group used less sow feed and all feed was less expensive per tonne. Labour and other costs were also the lowest of the four groups.

Herd sizes varied only slightly, as did age at weaning. Here again, the weight of weaners at eight weeks seemed to reflect the quantity of piglet feed consumed. The cost of piglet feed, together with labour and other costs, increased with total costs per weaner.

Tables 13.1 and 13.2 show that to produce low cost weaners it is essential to achieve a high output of weaners per sow. This comes from weaning early and then ensuring sows are properly served to produce at least 2.3 litters per sow in herd a year. A target of 11 born per litter should be the aim, and with low mortality it should be possible to produce 22 or 23 pigs per sow a year. Care to avoid wastage in the use of sow feed is necessary to contain costs. The sow feed, when shared by a high number, should not exceed 56 kg per weaner. Labour and other costs, which have risen substantially in recent years, need watching carefully. This does not mean low wages but ensuring that good stock people are employed, who are capable of attending to large numbers of sows efficiently so as to achieve maximum output. A highly paid person obtaining excellent results is likely to be less costly per pig produced than a low paid one achieving poor results.

Table 13.2 Breeding results related to the costs per weaner 1991

Range - costs per weaner	<£24	£24 - £26	£27 - £29	>£30
Number of herds	24	43	29	23
Number of sows in herd[(a)]	168	169	173	137
Litters per sow in herd	2.35	2.25	2.24	2.22
Age at weaning (days)	25	26	27	28
Live pigs born per litter	10.8	10.6	10.6	10.3
Weaners per litter at 8 wks	9.5	9.1	9.1	8.5
Weaners per sow in herd	22.2	20.5	20.3	18.8
Weight of weaners at 8 wks	18.0 kg	17.8 kg	18.5 kg	18.4 kg
Sow feed used per sow in herd	1.25 t	1.29 t	1.30 t	1.31 t
Sow feed used per weaner	56.3 kg	62.9 kg	63.9 kg	69.8 kg
Piglet feed used per weaner	17.2 kg	16.8 kg	17.8 kg	17.9 kg
Cost of sow feed per tonne	£144.70	£146.89	£149.18	£147.47
Cost of piglet feed per tonne	£253.46	£260.96	£273.78	£288.53
Compounds as % of total feed	67 %	66 %	81 %	62 %
Costs per weaner at 8 weeks	£	£	£	£
Feed	12.51	13.58	14.40	15.43
Labour	4.93	5.49	6.39	6.98
Other costs	4.97	5.43	6.40	7.55
Stock depreciation	.75	.91	1.33	1.83
Total breeding costs	23.16	25.41	28.52	31.79
Margin per £100 output	£9.47	£4.67	-£0.63	-£9.08

(a) Monthly average (including in-pig gilts).

Feeding herd costs

A similar grouping of herds in the feeding stage of production should also reveal which are the most important factors affecting profitability. Here, the groups have been selected on production costs per kg liveweight gain and those with the lowest costs should achieve the most efficient results in the key factors. The lowest cost group had total feeding costs at less than 48p per kg, while those of the highest group were 58p and over (Table 13.3).

Table 13.3 Feeding results related to the total costs per kg liveweight gain 1991

Range - costs per kg	<48p	48 - 52p	53 - 57p	>58
Number of herds	21	35	33	24
Feeder on hand per herd	871	877	1,035	599
Pigs produced per herd	3,212	3,343	3,661	2,426
Turnover per year	3.7	3.8	3.5	4.0
Liveweight of pigs produced	86 kg	87 kg	84 kg	82 kg
Liveweight of pigs brought in	17 kg	22 kg	20 kg	27 kg
Liveweight gain per pig	69 kg	65 kg	64 kg	55 kg
Pigs brought in - % purchased	15 %	35 %	26 %	56 %
Daily liveweight gain	.68 kg	.67 kg	.61 kg	.59 kg
Mortality percentage	2.2 %	2.2 %	2.3 %	2.6 %
Feed conversion rate	2.41	2.67	2.72	3.02
Cost of meal per tonne	£150.80	£146.35	£157.16	£162.60
Cost of feed per tonne[a]	£147.13	£140.61	£153.26	£162.32
Compounds as % of total feed	9 %	14 %	46 %	89 %
Other feed as % of total feed	5 %	7 %	5 %	1 %
Costs per kg liveweight gain	p	p	p	p
Feed	35.5	37.6	41.7	49.0
Labour	4.2	5.5	5.6	6.1
Other costs	5.1	6.1	6.8	7.2
Mortality	.8	1.0	1.1	1.6
Total feeding costs	45.6	50.2	55.2	63.9
Margin per £100 output	£10.36	£4.92	-£0.38	-£15.07

(a) Includes other feed (mainly by-products) "converted" to meal equivalent.

The key factor in feeding herd performance is the feed cost per kg liveweight gain, which is determined by the conversion rate and cost per tonne. Most of the other factors given in Table 13.3 have some influence on feed conversion rates and subsequently the costs per kg liveweight gain. Because conversion rates worsen as pigs increase in size, the weights of pigs brought in and produced are important. Light weights at starting and finishing should enhance conversion rates, while heavier weights depress them. The mixture of light and heavy weights in this comparison may need some interpretation, but the size of the difference in conversion rates between the groups clearly illustrates that they rise as costs per kg liveweight gain increases. Weights of pigs in and out of the herd also influence turnover, though this will be reduced if accommodation is not kept to capacity.

The group with total feeding costs of less than 48p per kg had the best feed conversion rate and the second lowest cost per tonne of feed, only 9 per cent of which were compounds. Feed costs per kg liveweight gain averaged 35.5p, considerably less than the other groups. Labour and other costs was also much lower.

The highest cost group not only had the worst conversion rate but also used the most expensive feed. Daily liveweight gain and the mortality rate were worse than for other groups, though more pigs were purchased - hence the higher weight of pigs brought in. The poor conversion and high cost feed gave an average feed cost of 49p per kg. With high labour and other costs, the total amounted to 63.9p per kg, 40 per cent more than the lowest cost group.

Price of finished pigs sold

Prices paid for finished pigs depend mainly on their quality and weight but also on finding a buyer paying good prices. Grading schemes to measure quality vary between buyers and it is difficult to produce any meaningful standards with which to make comparisons. Most buyers seek pigs in clearly defined weight bands and penalties for sending outside the required band can be severe. The smallest pigs usually make the highest prices per kg deadweight. Prices then tend to decline as the weight of pig increases, although pigs sold as baconers often make better prices per kg than cutters at slightly lower weights. In recent years the price differential between weight bands has narrowed considerably, often to only two or three pence per kg, whereas a decade ago differences of six or seven pence were common. Prices paid by buyers for pigs of similar weight seldom vary by much in the long-term, otherwise the lowest payers would not be offered many pigs, but prices in the short-term often differ according to supply and demand. If a buyer's trade increases, more pigs are required and those not on contract can usually be tempted away from an existing buyer with higher prices. The reverse sometimes happens: when a buyer's trade decreases and too many pigs are offered a reduction in price can cause some producers to seek an alternative outlet elsewhere. Once supply and demand are in balance, prices normally fall back into line. Reliable and prompt payments are other considerations that producers should look for in buyers.

Prices paid for pigs rarely take costs of production into account. For equal weight and quality, producers with high costs are not paid any more for their pigs than producers with low costs. Prices do vary between producers because of differing weight bands and quality (grading), not because of costs of production. Understanding the buyer's requirements is of paramount importance; then having the ability to supply suitable pigs to meet those requirements should ensure reasonable prices. Charges for marketing reduce prices. Most are on a per pig basis but haulage costs depend on the distance involved and the size of load. A full load usually works out much less per pig than a part load.

A comparison of results according to the average pig price received per kg deadweight is given in Table 13.4.

The group which received the lowest price of less than 96p per kg deadweight produced on average the heaviest pigs. Prices increased for the other groups as the weight per pig fell, except for the two highest price groups, which were both 62 kg. As expected, the group producing the heaviest pigs had a comparatively poor conversion rate but the feed was inexpensive. The group with the highest pig price also had a poor conversion rate with lighter pigs and dearer feed. Feed prices per tonne and costs per kg followed no regular pattern, though the group with the highest pig price per kg did have marginally higher costs per kg. This group produced more baconers than others and this would have contributed to the higher price.

Table 13.4 Feeding results related to the price per kg deadweight of pigs sold 1991

Range - pig price per kg	<96p	96 - 98p	99 - 101p	>102
Number of herds	24	33	33	21
Pigs produced per herd	3,126	2,672	3,987	3,144
Deadweight of sales	69 kg	65 kg	62 kg	62 kg
Net price per kg dwt[a]	93.6 p	97.5 p	100.4 p	103.6 p
Daily liveweight gain	.67 kg	.65 kg	.64 kg	.61 kg
Mortality percentage	2.4 %	2.1 %	2.4 %	2.2%
Feed conversion rate	2.76	2.62	2.64	2.75
Cost of meal per tonne	£149.30	£156.19	£151.80	£158.10
Cost of feed per tonne[b]	£140.54	£155.67	£149.71	£151.81
Compounds as % of total feed	28 %	49 %	29 %	42 %
Costs per kg liveweight gain	p	p	p	p
Feed	38.8	40.8	39.5	41.8
Labour	5.0	4.9	5.5	5.9
Other costs	6.3	6.3	6.3	6.2
Mortality charge	1.1	1.0	1.1	1.1
Total feeding costs	51.2	53.0	52.4	55.0
Sales distribution	%	%	%	%
Weaners/stores	1.0	3.7	1.9	3.4
Porkers	5.8	14.8	32.1	26.6
Cutters	61.1	45.5	55.7	17.2
Baconers	31.4	34.2	10.1	52.6
Heavies	.5	-	-	-
Casualties	.2	1.8	.2	.2
	100	100	100	100
Margin per £100 output	£2.65	-£0.43	£2.81	£2.14

(a) Haulage, marketing charges and levies have been deducted
(b) Includes other feeds (mainly by-products) "converted" to meal equivalent

Costs versus receipts

The best profit margins were achieved by the herds with the lowest costs of production. A high price per kg for finished pigs was helpful, especially when related to a heavier pig, but the overall price received accounted for little of the variation in profitability between herds. The level of costs was determined largely by the standard of performance achieved in the key factors of weaners per sow and feed requirements. As costs increased, profit margins fell, whereas a rise in pig prices, even after allowing for variation in weights of pig production, did not necessarily mean more profit. Often the advantage of higher prices was outweighed by the influence of production costs on profitability. Low costs almost certainly ensure good profit margins but high prices do not.

Breeding results by herd size

Comparing results by size of herd should indicate whether there are economies of scale in production costs, and such comparisons have been undertaken periodically for many years. Although herd sizes have increased, results seldom provide any conclusive evidence that larger herds have lower costs than smaller or medium sized herds. The 1991 results are typical of previous years.

As usual, size of herd seems to have little effect on the results in Table 13.5. The group of herds with more than 300 sows did have slightly lower costs per weaner but they were smaller weaners. If allowance were made for this difference in weight, costs per weaner would be much closer to the other groups. Mortality per litter (the difference between pigs born and weaners per litter) was lower for the largest herds at 1.3 deaths per litter, compared with 1.6 and 1.8 for the other two groups. The number of weaners per litter was virtually the same for the three groups, but the largest herds, with slightly more litters per sow in herd a year, produced more weaners per sow.

The group of largest herds used more sow feed per sow, probably because the proportion of outdoor units was greater, but when shared among the higher number of weaners produced, the quantity per weaner was little different from the other groups. The low requirement of piglet feed per weaner is accounted for by the lighter weight of weaners. The cost of feed per tonne for the three groups was remarkably similar for both sows and piglets.

Table 13.5 Breeding results related to herd size 1991

Number of sows in herd[a]	<100	100-300	>300
Number of herds	44	62	13
Number of sows in herd[a]	67	177	426
Litters per sow in herd	2.23	2.25	2.31
Age at weaning (days)	28	25	24
Live pigs born per litter	10.8	10.7	10.4
Weaners per litter	9.0	9.1	9.1
Weaners per sow in herd	20.2	20.5	21.0
Weight of weaners at 8 wks	18.3 kg	18.4 kg	17.4 kg
Sow feed used per sow in herd	1.25 t	1.28 t	1.32 t
Sow feed used per weaner	61.9 kg	62.6 kg	63.1 kg
Piglet feed used per weaner	17.8 kg	17.6 kg	16.6 kg
Cost of sow feed per tonne	£146.75	£146.80	£148.00
Cost of piglet feed per tonne	£270.01	£265.95	£267.00
Compounds as % of total feed	52 %	73 %	72 %
Costs per weaner at 8 wks	£	£	£
Feed	13.89	13.86	13.71
Labour	6.04	5.82	5.68
Other costs	5.59	6.00	5.84
Stock depreciation	.99	1.25	.93
Total breeding costs	26.51	26.93	26.16

(a) Monthly average (including in-pig gilts).

Costs per weaner at eight weeks were also little different, after allowing for the lighter weaners for the largest herds. Labour costs did however decline as the herds increased in size.

Feeding results by herd size

Similar comparisons relating to herd size were also undertaken for feeding stock, and results are given in Table 13.6

Table 13.6 Feeding results related to herd size 1991

Number of pigs in herd	<500	500-1,000	>1,000
Number of herds	38	40	35
Number of pigs in herd	308	712	1,638
Number of pigs produced per herd	1,275	2,789	5,814
Liveweight of pigs produced	80 kg	84 kg	87 kg
Liveweight of pigs brought in	22 kg	23 kg	20 kg
Pigs brought in - % purchased	27 %	35 %	31 %
Daily liveweight gain	.63 kg	.65 kg	.64 kg
Mortality percentage	2.2 %	2.0 %	2.5 %
Feed conversion rate	2.61	2.67	2.70
Cost of meal per tonne	£154.91	£153.74	£152.92
Cost of feed per tonne[a]	£154.59	£153.46	£146.35
Compounds as % of total feed	35 %	48 %	29 %
Costs per kg liveweight gain	p	p	p
Feed	40.3	41.0	39.6
Labour	5.9	5.4	5.2
Other costs	6.0	5.8	6.6
Mortality charge	1.1	1.0	1.1
Total feed costs	53.3	53.2	52.5

(a) Includes other feeds (mainly by-products) "converted" to meal equivalent.

As with the breeding results, the size of feeder herd seems to have little effect on costs of production. The largest herds produced the heaviest pigs and had a marginally worse mortality rate. Feed conversion rates increased with the weight of pigs produced, while the cost of meal per tonne declined slightly. The cost of feed was some £7 to £8 per tonne less for the group of largest herds, mainly because more use was made of cheaper by-products which formed some 8 per cent of the total. The use of by-products in the two smaller herd size groups was negligible.

The largest herds with their cheaper feed and only slightly worse conversion rate achieved the lowest feed cost per kg liveweight gain. Part of this was offset by higher other costs. Labour costs per kg liveweight gain fell as the size of herd increased. Total feeding costs were fractionally less per kg for the largest herds but the two groups of smaller herds were virtually the same.

In most years, herds grouped according to number have shown no clear economies of size. Perhaps this is not surprising when considering the many factors which influence production efficiency. The standard of housing, the health of the pigs, the suitability and cost of feed, together with the quality of stockmanship, are all of prime importance and apply equally to small and large herds. One would expect large units to buy feed more cheaply and there was some evidence of this with feeding stock but not in breeding herds. To be precise, the largest herds did show marginally lower costs in 1991 but, after allowing for weight variation, their savings were quite small, well under £1 a pig for breeding and feeding combined.

Results by farm size

Size of herd does not necessarily relate to size of farm. Often large herds are to be found on small farms and small herds on large farms, and there are many combinations of all sizes. A study of pig results by farm size may provide some evidence of economies of scale. For several years herds in the scheme have been compared according to the size of farm business in which they are embedded, measured in terms of British Size Units (BSU), by EC standard methodology. Average results over the five year period 1987-91 are shown in Table 13.7 for four different size groups.

A five year period has been chosen to provide a more stable basis for this comparison. The pig units within small farm businesses of under 16 BSUs a year have clearly done less well than those within medium and large businesses. The pigs' margin per £100 output was considerably lower because of much higher breeding costs per weaner and feeding costs per kg liveweight gain. Both sow and piglet feeds per tonne were, on

Table 13.7 Results by farm size (EC standard units)
(five year average 1987-91)

Farm business size(a) in terms of BSUs	<16	16 - 40	40 - 100	>100
Number of herds	22	37	43	46
Total farm area (ha)	8.6	36.0	102.5	484.4
Area of crops (ha)	5.2	30.1	90.1	363.5
Composition of total SGM	%	%	%	%
Cereals	14.5	29.4	34.6	41.0
Sugar beet & potatoes	2.5	8.1	9.1	16.9
Other arable	.9	4.3	8.7	9.9
Horticulture	.1	.5	3.0	6.5
Pigs	81.0	54.0	40.1	19.9
Other livestock	1.0	3.7	4.5	5.8
	100	100	100	100
Pigs' margin per £100 output(b)	£1.96	£8.02	£10.33	£9.15
Breeding				
Number of sows in herd(c)	51	98	145	186
Litters per sow	2.26	2.28	2.26	2.26
Live pigs born per litter	10.8	10.8	10.6	10.6
Weaners per litter at 8 wks	9.1	9.3	9.1	9.3
Weaners per sow in herd	20.7	21.1	20.6	20.9
Sow feed used per weaner	61.6 kg	59.2 kg	60.9 kg	62.3 kg
Piglet feed used per weaner	17.7 kg	17.1 kg	18.0 kg	17.8 kg
Cost of sow feed per tonne	£147.19	£145.87	£142.24	£141.10
Cost of piglet feed per tonne	£257.43	£255.85	£247.34	£254.23
Compounds as % of total feed	90 %	76 %	63 %	65 %
Feed costs per 8 wk weaner	£13.61	£12.99	£13.11	£13.29
Total costs per 8 wk weaner	£25.54	£23.78	£23.67	£24.53
Feeding				
Liveweight of pigs produced	66 kg	53 kg	68 kg	77 kg
Mortality percentage	1.3 %	1.0 %	1.7 %	2.0 %
Feed conversion rate	2.85	2.54	2.71	2.70
Cost of meal per tonne(d)	£157.91	£161.98	£151.94	£150.51
Cost of feed per tonne(d)	£157.76	£161.71	£148.63	£146.90
Compounds as % of total feed	77 %	53 %	37 %	34 %
Feed costs per kg lwt gain	45.0 p	41.0 p	40.2 p	39.6 p
Total costs per kg lwt gain	55.6 p	50.9 p	50.7 p	50.6 p
Sales distribution	%	%	%	%
Weaners/stores	34.7	52.4	31.4	18.4
Porkers	19.1	18.1	20.1	15.5
Cutters	45.8	20.2	25.6	36.0
Baconers	.2	8.4	20.6	26.9
Heavies	-	.6	2.1	2.8
Casualties	.2	.3	.2	.4
	100	100	100	100

(a)   Farm business size in the UK is measured in British Size Units (BSU) with one BSU equal to 2,000 ECUs of Standard Gross Margin at 1978-80 values
(b)   No charge has been included for interest on capital
(c)   Monthly average (including in-pig gilts)
(d)   Includes other feed (mainly by-products) "converted" to meal equivalent

average, the most expensive; only the next size group had higher costs for with feeding stock.

With limited farm area the small farm businesses had little opportunity to make money at anything apart from pigs, which formed 81 per cent of the total SGM. Their comparatively expensive feed was obviously a handicap. Breeding performance in these smaller herds was fairly similar to other groups but feeding stock results were depressed by a poor conversion rate. Labour and other costs (the difference between feed costs and total costs) were highest of the breeding herds though more in line with others for feeding stock. Nearly 70 per cent of this labour on pigs was farmers' own or family which, although charged for at standard rates, was not actual payments and may have helped solvency in difficult times.

By contrast the farmers' own or family labour on pigs for the group of large farm businesses of over 100 BSUs a year was less than 10 per cent of the total. Two-thirds of the total SGM came from arable crops, plus another 6 per cent from horticulture. Their pig units were much larger on average than other groups but formed only 20 per cent of the total SGM. The pigs' margin per £100 was similar to the two medium sized business groups, as was performance in most production factors. The main advantage was from comparatively low cost feed for sows and feeding stock.

Results by degree of specialisation

Another method of comparing results is by the importance of pigs in relation to other farming activities. Farms were rated according to the percentage of total SGM attributed to pigs; Table 13.8. Those with more than 66 per cent were classed as highly specialised, 33 to 66 per cent as medium and under 33 per cent as low. In other words the highly specialised group had only minor distractions elsewhere and were able to concentrate their attention on pigs, whereas those of low specialisation had other demands on their time and occasionally management of the pig unit must have received a low priority. Usually the highly specialised farms had only small areas to grow crops and on average over the five years, 1987-91, pigs formed 91 per cent of the total SGM of the farms in this group. For the medium group with more land, pigs contributed 45 per cent of total SGM. The group

Table 13.8 Results by degree of specialisation
(five years average 1987-91)

Pigs as % of total farm SGM[a]	High > 66%	Medium 33 - 66%	Low < 33%
Number of herds	43	31	74
Total farm area	14.3 ha	103.8 ha	322.7 ha
Area of crops	9.4 ha	88.3 ha	246.6 ha
Composition of total SGM	%	%	%
Cereals	6.0	32.0	45.3
Sugar beet and potatoes	1.5	9.3	16.4
Other arable	.6	6.0	10.4
Horticulture	.1	3.5	5.0
Pigs	91.1	44.9	16.6
Other livestock	.7	4.3	6.3
	100	100	100
Pigs' margin per £100 output[b]	£8.54	£9.54	£8.63
Breeding			
Number of sows in herd[c]	163	156	106
Litters per sow	2.29	2.25	2.23
Live pigs born per litter	10.7	10.6	10.7
Weaners per litter at 8 wks	9.2	9.2	9.2
Weaners per sow in herd	21.1	20.7	20.7
Sow feed used per weaner	60.2 kg	60.8 kg	62.7 kg
Piglet feed used per weaner	17.3 kg	17.8 kg	17.8 kg
Cost of sow feed per tonne	£144.62	£142.26	£141.26
Cost of piglet feed per tonne	£253.80	£255.89	£250.03
Compounds as % of total feed	83 %	60 %	60 %
Feed costs per 8 wk weaner	£13.08	£13.20	£13.27
Total costs per 8 wk weaner	£23.82	£24.11	£24.40
Feeding			
Liveweight of pigs produced	57 kg	70 kg	77 kg
Mortality percentage	1.4 %	1.6 %	2.0 %
Feed conversion	2.64	2.69	2.72
Cost of meal per tonne	£160.44	£151.02	£150.52
Cost of feed per tonne[d]	£158.20	£147.59	£147.41
Compounds as % of total feed	66 %	38 %	31 %
Feed costs per kg lwt gain	41.7 p	39.6 p	40.0 p
Total costs per kg lwt gain	51.7 p	50.0 p	51.2 p
Sales distribution	%	%	%
Weaners/stores	47.2	28.3	18.1
Porkers	18.3	17.7	17.2
Cutters	24.4	32.8	36.1
Baconers	9.5	18.9	25.2
Heavies	.4	1.8	3.1
Casualties	.2	.5	.3
	100	100	100

(a) Standard Gross Margin by EC typology
(b) No charge has been included for interest on capital
(c) Monthly average (including in-pig gilts)
(d) Includes other feed (mainly by-product) "converted" to meal equivalent

of low specialisation was mainly arable farmers with pig units of varying sizes but small (17 per cent) in relation to their total farm businesses.

Farm sizes in terms of area of the three groups were vastly different. The highly specialised pig farms averaged 14 hectares, compared with 103 hectares for the medium group and 322 for low specialisation. Herd size was largest for the high specialists with an average 163 sows per herd, closely followed by 156 sows for medium and a much smaller 106 for low. The margin per £100 output from pigs was similar for the three groups, as also was breeding performance in numbers of litters and weaners produced. Slightly more sow feed was used by the group of low specialisation but it was less expensive and brought both feed and total costs per weaner nearer back into line with the two other groups.

With nearly half of the output of pigs sold as weaners or stores, the average weight of pigs produced for the high specialists was much lower than the other two groups. This variation in weight of pigs produced in the feeding section largely accounts for the difference in feed conversion rates, which followed a normal pattern of rising as the weight of pigs increased. Feed cost more per tonne for the high specialist group, making feed costs per kg liveweight gain higher than others, but this was partly offset by lower labour and other costs to bring total costs per kg liveweight gain closer to the other groups.

Chapter 14

Capital Requirements

Of the many methods of assessing profitability one of the best is to compare the returns on capital investment. This is easy to state as a proposition but often difficult to translate into practice. One problem is to arrive at the amount of capital invested in pigs by individual farmers on a basis which allows comparison to be made. Calculations of capital required to start new units on green field sites are fairly straightforward. More difficult are assessments of the amount involved in existing units.

The range and value of housing and equipment in use is vast. Some farmers have erected buildings at recent high prices, while others have buildings twenty or more years old which date from a period when prices were much lower. The question arises whether the latter should be valued at original cost less depreciation (probably a very small figure now) or at cost of replacement (a much higher one). Indeed some feel that buildings should be written-off over a given period, say ten years, and their subsequent value ignored. This overlooks the fact that they do still have a value, which can sometimes be realised by renting out buildings no longer in use to other pig producers.

Even if buildings can be valued, the capital investment is affected by the particular circumstances of each farm. An owner-occupier has a substantial sum invested in buildings; a tenant who rents buildings from his landlord may have none. Many pig units have expanded over the years from small beginnings; some may have retained home-bred gilts for breeding and financed new buildings and equipment from previous profits. Own mixed feed is usually cheaper per tonne but more capital will be required for plant and storage than with purchased compounds. Furthermore, the amount of working capital required for feed will depend on whether it is paid for in advance, in the month following delivery, or still further in arrears, a practice which may then entail interest charges. Stocking density, throughput and type of production are some other considerations involved. It is obvious, therefore, that in comparing return on capital, certain standard conditions must be assumed for each type of production but for individual units the prevailing circumstances can change the result considerably.

Established herds

In established herds the average amount of capital involved is likely to be less than that needed to start a new unit with purchased breeding stock, new buildings and equipment at present day prices. The following assessments for the established herds in the Cambridge scheme assume standard conditions for each type of production. Breeding stock were valued at the average of the opening and closing valuations. For feeding-only herds, the average price of pigs purchased during the year was used. All buildings were included at estimated current value for both owned and rented farms. Rented buildings away from the farm were capitalised at ten times the annual rent. Major machinery used mainly for pigs such as tractors, vehicles, balers, bale handling equipment, milling and mixing plants were included at current values for that proportion due to pigs. Other equipment was included at its depreciated value. Working capital for feed, labour and other costs was assessed on an even throughput of pigs throughout the year. For breeding stock, working capital covered the average breeding cycle (23 weeks in 1991) and was extended by the length of time the progeny remained on the farm until sale, thus releasing the working capital to finance the next batch. In established herds, where farrowings are evenly spread throughout the year, the average capital per litter for breeding stock will at any one time be approximately half of the total cost. Similarly for feeding stock with spread production, the average working capital per pig would be about half the cost of rearing it. With batch production, operating an all in, all out system, working capital would equal the total costs of each batch.

By system

The average capital requirements and returns on capital for the three systems of pig keeping are given in Table 14.1. Assessments for the breeder-feeder herds and the mainly breeding herds selling weaners and stores are on a "per sow" basis, while the feeding-only herds are "per pig".

Table 14.1 Average capital requirements and returns by system 1991

System	Breeder-feeders per sow	Mainly breeding per sow	Feeding-only per pig
Number of herds	82	37	31
Capital requirements per sow	£	£	£
Value per sow	132	134	-
Share of boar's value	17	18	-
Cost of weaner	-	-	32.79
Buildings and equipment	694	415	73.19
Working capital	264	151	15.65
Total capital	1,107	718	121.63
Margin per sow a year[a]	£30.10	£3.76	-£1.66[b]
Return on capital	%	%	%
1991	2.7	.5	-1.4
1990	30.7	29.0	21.3
1989	15.7	11.2	22.4
1988	-3.6	-1.7	.7
1987	10.9	5.7	8.5

(a) No charge has been included for interest on capital
(b) Margin per pig place

Capital requirements were, as expected, much higher for the breeding and feeding herds where the progeny were retained until reaching slaughter weight. The mainly breeding herds, selling weaners at about 10 to 12 weeks of age, saved capital on buildings and equipment by not finishing pigs and their working capital was also correspondingly lower. Accommodation from eight weeks of age had to be provided for some five to six pigs per sow for the breeder-feeder herds, whereas on average only one or two weaners per sow remained in the mainly breeding herds after eight weeks. Pigs in the feeding-only herds came in heavier and, therefore, occupied housing for less time. Most managed a turnover rate of nearly four batches a year.

The relative margins and returns on capital between mainly breeding herds and feeding-only depends largely on weaner prices. When they are high the breeders benefit but when they are low the feeders have the advantage. The breeder-feeders, being isolated from fluctuating weaner prices, usually have more stable margins and returns on capital. They normally produce weaners cheaper than the cost of buying them, avoiding the breeders' profit, and they save on haulage.

The contrast in returns on capital between 1990 and 1991 could hardly be greater. This was a reflection of pig prices which were high in 1990 and low in 1991. In 1989, the feeding-only herds benefited from a rapid rise in finished pig prices after having bought them cheaply as weaners a few months earlier.

Weaner production

In assessing capital requirements for the mainly breeding herds selling weaners a distinction between indoor and outdoor production is needed. Housing an outdoor unit requires less capital than an indoor one, though the amounts for stock and working capital are similar. With an outdoor unit there is always the vexed question of whether to include the value of land in capital. Often land is rented for these units and a charge has been included for both owned and rented land, as well as expenditure on fencing, water supplies, etc, but the value of the land, even if owned, has not been included in capital assessments. Two-thirds of the small sample of outdoor units sold weaners and most had relatively new huts, although often equipment was second-hand or had been transferred from arable use. In contrast, many of the buildings used by the indoor herds were much older, as poor profits from pigs in recent years had limited new investment. A study of indoor and outdoor herds is likely to compare mostly fairly new huts on outdoor units with older buildings for indoor herds.

**Table 14.2 A comparison of average capital requirements per sow
and returns for indoor and outdoor herds selling weaners 1991**

	Indoor	Outdoor
Number of herds	30	7
Capital requirements per sow	£	£
Value of sow	134	132
Share of boar's value	17	22
Buildings and equipment	463	256
Working capital	<u>132</u>	<u>127</u>
Total capital	<u>746</u>	<u>537</u> (a)
Margin per sow a year[b]	£3.41	£4.89
Return on capital	%	%
1991	.5	.9
1990	26.6	41.5
1989	7.6	24.2
1988	-3.0	5.8
1987	3.2	21.6

(a) Excludes capital value of land
(b) No charge has been included for interest on capital

There was little difference between the two groups in the value of sows or for working capital involved as shown in Table 14.2. The share of the boar's value per sow was higher for outdoor units as they averaged one boar to 15 sows compared with one to 19 sows for the indoor units and had fewer sows to share this value. The total capital for the outdoor units of £537 per sow was 28 per cent lower than the £746 for indoor herds. Buildings and equipment were 45 per cent lower.

Margins per sow for 1991 were depressed by low pig prices. Although returns on capital of 0.5 per cent for the indoor group and 0.9 per cent for outdoors appear similar, the latter was actually nearly double that for indoors. Returns on capital for the previous years, when different prices prevailed, provide more realistic comparisons. The outdoor units with lower production costs produced the best margins and their lower capital requirements show far superior returns on capital in each of the four previous years.

Finished pig production

For breeding and feeding herds finishing home-bred weaners and for feeding-only herds finishing purchased weaners the amount of capital required depends on the type of production undertaken. The larger pigs require more space and therefore fewer are accommodated in each of the feeding pens. They also occupy the pens longer so that a greater number are needed. Often some saving is possible in the number of pens required by doubling the number of small pigs per pen and taking half of them to other pens when they out-grow the original pen.

Finishing home-bred weaners to slaughter weights is an extension of breeding the weaners and the amount of capital required must cover both the breeding and feeding stages of production. The capital assessments and returns shown in Table 14.3 are therefore on a "per sow" basis and take into account the fact that larger pigs remain on the farm longer and require more feed and attention.

These assessments of capital were intended to reflect the average amounts required by the established herds in the scheme. The value per sow and share of the boar's value used in the three examples was the average of all breeder-feeder herds, on the assumption that all types of finishing start with a weaner, and using the average as standard focuses the comparison on the different types of production. Buildings and equipment values, as also working capital and total capital, increased with the size of pig produced and the time taken by the progeny to reach slaughter weight.

The returns on capital for 1991, like margins per pig, were quite low and contrasted markedly with 1990 when pig prices were much higher. On average over the five-year period 1987-91, the returns from breeding and feeding for cutter and baconer production were similar at 12.1 and 13.0 per cent respectively, while that for porkers (7.7 per cent) was much less. Individual years varied considerably, ranging from 1988, a bad year for pig producers, to 1990, a very good one.

Table 14.3 Average capital requirements per sow and returns for different types of production 1991 (breeding and feeding)

	Porkers	Cutters	Baconers
Capital requirements per sow	£	£	£
Value of sow[a]	132	132	132
Share of boar's value[a]	17	17	17
Buildings and equipment	648	720	763
Working capital	238	283	304
Total capital	1,035	1,152	1,216
Margin per sow a year[b]	-£12.51	£32.59	£62.20
Return on capital	%	%	%
1991	-1.2	2.8	5.1
1990	24.0	31.2	30.4
1989	12.0	16.3	16.5
1988	-4.3	.4	.8
1987	8.2	9.8	12.4

(a) Average of all breeding herds
(b) No charge has been included for interest on capital

For the feeding-only herds that relied on purchased weaners, capital assessments were on a per pig basis and included the investment in weaners at the average price paid for the year. When weaners are expensive this investment is correspondingly higher. Assessments of buildings and equipment values and working capital were on the same basis as for the breeding and feeding herds. All buildings were included at current values, whether used to capacity or not, and the total shared by the average number of feeder pigs over the year. In times of unprofitable production some farmers reduce pig numbers and this would partly account for higher buildings and equipment values per pig than in breeder-feeder herds. They also start with a heavier weaner which requires more space. Comparisons for the three types of production are given in Table 14.4.

**Table 14.4 Average capital requirements per pig and
returns for different types of production 1991
(feeding-only)**

	Porkers	Cutters	Baconers
Capital requirements per pig	£	£	£
Cost per weaner[a]	32.79	32.79	32.79
Buildings and equipment	58.55	71.73	76.12
Working capital	11.78	14.99	16.18
Total capital	103.12	119.51	125.09
Margin per pig place a year[b]	-£13.07	-£1.28	£4.22
Return on capital	%	%	%
1991	-12.7	-1.1	3.4
1990	8.7	25.1	23.3
1989	10.9	20.4	20.1
1988	-13.8	.4	.7
1987	5.5	9.7	15.1

(a) Average cost of weaners purchased by the feeding-only herds (28 kg)
(b) No charge has been included for interest on capital

As in previous comparisons, the amount of capital increases with the size of pig produced because they stay on the farm longer and use more feed. The returns on capital were very poor in 1991. Baconers performed best and were the only group to show a positive return. The two previous years, with higher pig prices and margins, were much better and among the best on record. Over the five-year period 1987-91, both cutters and baconers had reasonable returns on capital, averaging 10.9 and 12.5 per cent respectively before charging interest, levels fairly similar to those of breeding and feeding herds, whereas porkers only managed a break-even return on average over the five years.

Variation

It should be remembered that these comparisons of returns on capital are group averages and that individual herds results varied considerably around the average. For example, feeding-only porker producers have not been very successful as a group, but the best individual herds achieve returns as good as the average cutter and baconer herds. The much shorter finishing period for porkers from say 28 kg at purchase to 75 kg liveweight (55 kg deadweight) at sale offers little opportunity for the producer to make an impact even with efficient performance. There is a further handicap if the weaners bought in were expensive, as there is less weight gain over which to spread this overhead cost compared to the heavier cutters and baconers.

The variation from herd to herd within each group is usually greater than the difference between groups, indicating that the level of efficiency achieved is more important than systems or types of production. Table 14.5 shows details for the 20 most profitable herds in 1991.

Table 14.5 Systems and types of production for the 20 most profitable herds 1991

	Total	Weaners	Porkers	Cutters	Baconers	Mixed
Breeding & feeding	12	-	3	5	2	2
Mainly breeding	5	5	-	-	-	-
Feeding-only	3	-	-	1	2	-
Total	20	5	3	6	4	2

The 20 most profitable herds included a wide mix of systems and types of production. The 12 breeding and feeding herds constituted 14 per cent of all breeder-feeder herds, compared with 13 per cent for the five mainly breeding and 10 per cent for the three feeding-only. Cutters were produced by six of the 20 most profitable herds, weaners by five, baconers by four, porkers by three and two sold pigs in more than one category.

Comparative prices

Pig prices fluctuate from time to time and this can alter the order of profitability for different types of production. Most producers contract to sell their pigs, or the majority of them, for a year at a time. They usually choose an outlet on previous experience of prices paid and their ability to supply suitable pigs. For them marketing arrangements are fixed and only come up for review once a year. Other producers without contracts can be more flexible and they may have a choice of changing the size of finished pigs produced when prices are favourable. They are at risk, however, in times of surplus when pigs on contract have priority and those without may only be accepted at lower prices. It is apparent that total costs per kg deadweight fall as the size of pig

Table 14.6 Comparative pig prices per kg deadweight
to give equal returns on capital

50 kg	60 kg	70 kg	80 kg
p	p	p	p
104.9	98.5	93.6	89.7
107.5	100.8	95.8	91.8
110.0	103.2	98.0	93.9
112.6	105.6	100.2	96.0
115.2	107.9	102.4	98.1
117.7	110.3	104.6	100.1
120.3	112.6	106.8	102.2
122.9	115.0	109.0	104.3
125.4	117.4	111.2	106.4
128.0	119.7	113.4	108.5
130.6	122.1	115.6	110.5
133.2	124.5	117.8	112.6
135.7	126.8	120.0	114.7
138.3	129.2	122.2	116.8
140.9	131.5	124.5	118.9

increases and, therefore, small pigs need a higher price per kg to be as profitable as larger pigs. When prices vary for different types and sizes of pigs, producers not already committed by contract may wish to consider whether there are advantages from marketing their pigs at a different weight. From the scheme data, Table 14.6 has been prepared to illustrate the prices needed to give the same returns on capital over deadweights ranging from 50 to 80 kg. To ensure a sound comparative basis, performance was based on the latest five years' results and although calculated for breeding and feeding herds, the prices given should be equally applicable to feeding-only herds.

These comparisons are based on average costs and capital requirements and take into account space occupied and age at slaughter. They show, for example, that 120.3p per kg deadweight for pigs of 50 kg gives the same return on capital as 112.6p for pigs of 60 kg and 106.8p for pigs of 70 kg. A producer on average efficiency receiving 113p per kg deadweight for 60 kg pigs would be better off from an offer of 109p for 70 kg pigs. For the future, costs may vary but the relationship of these prices is unlikely to change by more than a penny or two per kg and should provide a guide to relative profitability for several years.

Price differentials per kg for pigs between 50 and 60 kg have declined considerably in recent years and few opportunities now exist to match the figures given in Table 14.6. Confirmation of this deterioration in the prices paid for lightweight pigs comes from comparisons of profitability between porkers and cutters, where the heavier cutter pigs now on average come out best. Unless the price differentials improve for small pigs, it will pay producers to carry them on to at least 60 kg, even if this means expenditure on alterations to pens to accommodate heavier pigs.

Table 14.7 Average sale weights for a constant sample of 48 herds

Year	Average deadweight
	kg
1981	62.0
1983	62.2
1985	63.7
1987	64.1
1989	65.2
1991	65.3

Changes in pricing structures have resulted in a steady increase in the average weight at sale for herds in the scheme. Details are given in Table 14.7.

Between 1981 and 1991 the average deadweight of pigs at sale increased by 3.3 kg, or just over 5 per cent. Only 12 of the 48 herds in the scheme continually during this period produced lighter pigs in 1991 than in 1981. Seven of them produced pigs marginally lighter by a kilo or two, and three were heavy pig producers in 1981 that had changed to cutters as fewer heavies were required. Most of the remaining 36 herds increased the weights of pigs sold during this period.

Chapter 15

Updating Costs and Returns

Fluctuations

Since the 1970s costs and returns of pig production have fluctuated frequently and often by substantial amounts. Subsequent profitability has often changed drastically in a short space of time. While field surveys are vital in establishing reliable costs and sound measures of physical performance, many in the pig industry seek more up to date financial facts than annually published results can provide. A method of updating the latest known annual results for changes in costs and returns helps to satisfy these needs.

By necessity annual results are produced in arrears and average costs and returns usually conceal wide variations which have occurred during the year. For example, the net price received for cutters for the recording year ended 30 September 1990 averaged 117p per kg deadweight but ranged from 100p in January to 137p in June. Costs for breeding and feeding herds averaged 91p per kg for the year but ranged from 95p in January (high costs at a time of low pig prices) to 89p in May, when pig prices were rising. The resulting margins averaged 26p per kg but varied between 6p in January to 47p in June.

Costs for feeding-only herds include the price of purchased weaner and these too varied reflecting the changes in finished pig prices. For 1990 their costs of production averaged 107p per kg deadweight but ranged from 111p in March, when weaners were expensive, to 98p in September, when they were comparatively cheap.

Pig prices, breeding and feeding costs and feeding-only costs are shown separately in Figure 15.1. Monthly averages can be compared with annual averages (the flat line) over the five-year period from October 1986 to September 1991. These costs and

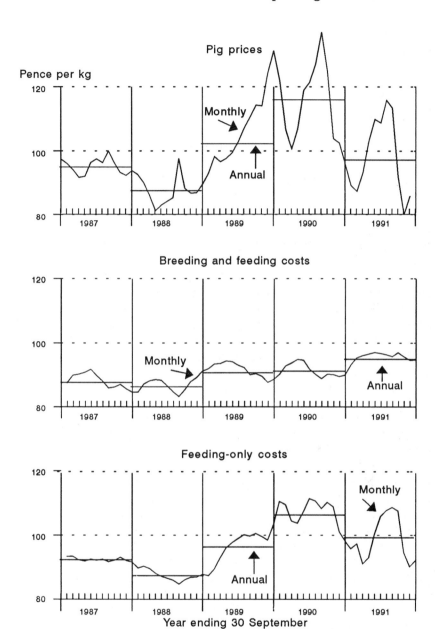

**Figure 15.1 Monthly costs and prices per kg deadweight
for cutter production related to annual averages**

returns relate to cutter pigs, though similar changes occurred to porkers and baconers.

The monthly costs and prices differed considerably from the annual averages. Pig prices were especially volatile, varying by at least 35p per kg deadweight between the highest and lowest monthly average in each of the last three years. Costs for feeding-only herds, with their purchased weaners, also changed drastically during this period, ranging from 13p to 18p difference between the highs and lows for each year. By comparison costs for breeding and feeding herds were more stable, fluctuating only by between 4p and 7p per kg each year. Production costs were reasonably stable between 1985 and 1991 but previously they had risen almost continuously for 12 years, mainly due to escalating feed prices. Future uncertainties of cereal prices could bring about more variation in production costs in the years ahead.

As a management aid it is helpful to keep up to date with costs and returns from frequently changing prices. Some need this information for their own business; others to be aware of the effect of changing situations in connection with a service they provide for pig producers. Methods of undertaking this work are described in the following pages.

Data base

The essential ingredients of a regular monitoring of the profitability of pig production are as follows:

1. A reliable knowledge of representative performance achieved by commercial pig units, together with details of quantities and kind of feed used, labour and other costs including stock depreciation and mortality charges for the different types of production
2. A sound basis for monitoring changes in production efficiency, especially those affecting feed requirements
3. A method of updating costs promptly from dependable sources (because of the extent of price changes nowadays, a high degree of accuracy is important)
4. The ability to verify the results

Table 15.1 Data bank for 1992 updating

| | Breeding Weaners[a] 8 wks | Feeding from 8 wks[c] | | |
		Porkers 55 kg dwt	Cutters 66 kg dwt	Baconers 69 kg dwt
Costs per pig	£	£	£	£
Feed	13.99	22.74	27.79	29.55
Labour	5.73	3.03	3.65	3.78
Other costs	5.98	3.32	4.22	4.60
Stock depreciation	1.29	-	-	-
Quantity of feed	81 kg	143 kg	185 kg	200 kg
Cost of feed per tonne	£172.78	£158.67	£150.14	£147.80
Mortality percentage[d]		2.5 %	2.5 %	2.5 %

| | Breeding Weaners[b] 30 kg | Feeding from 30 kg[e] | | |
		Porkers 55 kg dwt	Cutters 66 kg dwt	Baconers 69 kg dwt
Costs per pig	£	£	£	£
Feed	18.18	19.14	24.23	26.00
Labour	6.53	2.60	3.26	3.41
Other costs	6.37	3.22	4.16	4.56
Stock depreciation	1.29	-	-	-
Quantity of feed	103 kg	120 kg	162 kg	177 kg
Cost of feed per tonne	£176.53	£154.08	£145.78	£143.59
Mortality percentage[d]		1.8 %	1.8 %	1.8 %

(a) Breeding and feeding herds
(b) Breeding herds selling weaners
(c) From eight weeks (18 kg)
(d) Standardised for all types
(e) Feeding-only herds

Collecting this information to achieve an acceptable standard can be rather time consuming but this is no doubt worthwhile providing there is confidence in the results. A survey of representative pig units, such as the Cambridge scheme, where detailed records of stock and feed are regularly collected and scrutinised for feasibility, should provide a suitable basis for this purpose. The most recent annual results form the data base for the updating and can be replaced each year as new results become available. The 1991 average results given in Table 15.1 provide the data bank for 1992 monthly updating.

Costs and feed requirements increased with the size of pig produced. The feed was more expensive per tonne for breeding herds selling weaners than for the breeding and feeding herds to eight weeks of age, where the progeny were retained for finishing to slaughter. In the feeding stage the cost of feed per tonne declined as the weight of pig produced increased for both herds finishing their home-bred weaners and for feeding-only herds using purchased weaners. Although actual mortality varies by type of production and from year to year, average rates have been used here for simplicity. For total costs per pig for breeding and feeding, the figures under weaners to eight weeks and feeding from eight weeks should be added together according to type.

As feed costs per tonne varied so too did the type of feed used. The main ingredients used in 1991 are listed in Table 15.2.

Table 15.2 Feed data base

| Feed[a] | Breeding | | Feeding | | |
	Weaners 8 wks	Weaners 30 kg	Porkers	Cutters	Baconers
	%	%	%	%	%
Breeder compounds	41.1	53.0	-	-	-
Piglet compounds	17.6	34.8	-	-	-
Feeder compounds	-	-	46.0	41.4	28.9
Breeder conc mixes	2.3	1.3	-	-	-
Feeder conc mixes	-	-	7.2	2.7	2.6
Wheat	17.2	5.6	21.8	22.5	29.3
Barley	10.9	2.9	15.0	13.1	15.5
Milling offals	6.0	1.0	3.9	5.4	6.8
Soya bean	3.1	.9	5.4	12.6	14.9
Fish meal	1.8	.5	.7	2.3	2.0
	100	100	100	100	100

(a) The list of feeds used has been condensed (and percentages marginally adjusted) to correspond with feed quotations readily available.

Some other ingredients used have been amalgamated with similar types but small quantities and feed additives have been omitted. The list given was usually sufficient to measure cost changes satisfactorily and represents the feeds used by the herds in question to achieve the performance shown in Table 15.1.

Updating costs

To update costs promptly it was necessary to find a source of reliable price quotations other than from members of the scheme. While most co-operators sent in their records during the month following that to which they applied, some waited longer and delayed the availability of data suitable for updating purposes. Anyway some details were only collected at the end of each six month period. Alternative methods of collecting prices had to be found if the results were to be produced quickly. To update feed costs, prices were tabulated from several sources including compounders, MAFF, HGCA and the farming press. The latest quotations were related to the average of the same quotations over the year of the scheme and the difference applied to last year's feed costs according to the quantities used for each type of production. Using the difference between the previous year's and the latest quotations to arrive at current costs largely overcomes problems associated with discounts and transport charges. An example of updating feed costs for producing eight week weaners for the month of March 1992 is given in Table 15.3.

The example shows that the average cost of feed for producing eight week weaners rose by £5.65 per tonne in March from the previous year's level. For the average quantity of 81 kg, feed costs per weaner had increased by 46p from £13.99 for the previous year to £14.45 for March.

Labour costs were updated according to changes in wage rates and national insurance contributions. An increase in other costs was allowed for changes in veterinary bills, electricity and water charges, etc. Also, a mortality charge should be included to allow for the cost of weaners that die after entering the feeding herd. The mortality charge is calculated by using the average production costs for home-bred weaners from breeder-feeder herds, or the purchased price for bought in weaners for feeding-only herds, multiplied by the mortality rate.

Table 15.3 An example of updating feed costs for producing weaners (8 weeks)

	Average quotations for previous year (1991)	Quotations for March 1992	Change per tonne	% used	Value per tonne
	£	£	£		£
Breeder compounds	164.86	171.45	+6.59	41.1	2.71
Piglet compounds	272.62	279.40	+6.78	17.6	1.19
Concentrate mixes	276.39	283.96	+7.57	2.3	.17
Wheat	115.25	117.77	+2.52	17.2	.43
Barley	107.17	109.12	+1.95	10.9	.21
Milling offals	105.60	108.10	+2.50	6.0	.15
Soya bean	136.52	150.50	+13.98	3.1	.43
Fish meal	334.90	355.00	+20.10	1.8	.36
					5.65

Seasonal variation

Annual costs were used for the data base but performance and production costs vary throughout the year according to season. More pigs are reared per sow in summer than winter and less feed is required for both breeding and feeding stages of production. To provide realistic assessments of costs each month, the updated monthly costs need adjusting to take account of this variation. Results over several years show that costs of production are on average approximately 3 per cent above the annual level in the winter and 3 per cent below in the summer. These adjustment rates have been used for four of the six months of winter and summer. The percentage adjustments for the intermediate months have been graduated to provide a smooth change from one season to the next. Over the twelve months the following adjustments to costs have been applied:

Month	Jan	Feb	Mar	Apr	May	Jun	Jul	Aug	Sep	Oct	Nov	Dec
% of	%	%	%	%	%	%	%	%	%	%	%	%
costs	103	103	101	99	97	97	97	97	99	101	103	103

The example given earlier of updated feed costs for producing eight week weaners of £14.45 per pig for March would now be seasonally adjusted at the appropriate rate of 101 per cent to give £14.59.

Production cycle

There are two main methods of updating costs:

1. spot - estimated costs at current prices
2. spread - estimated costs covering the production period

It is important to distinguish between the two, as the difference can be quite substantial in times of rapidly changing prices. Spot represents current costs at prices prevailing at the time of slaughter and spread costs cover the production cycle of the pigs just slaughtered. For breeding and feeding the progeny to cutter weight (say 66 kg deadweight), the spread production period amounts to some nine or ten months - from the time the sow is served, through gestation and rearing, to slaughter. For a feeding-only unit finishing purchased weaners for cutters, the spread production period will be about three months. Costs are allocated over the production period on the basis that, apart from gestation in breeding herds, they rise each month as the weight of pig increases and the latter months receive a higher proportion of the total. When prices are rising, spot costs will be higher than those spread over the production period. Should prices fall, then spread costs will for a time become higher than spot. Both methods of updating serve useful purposes, depending on whether costs of producing the pigs just slaughtered are required, or the present day costs at current prices.

Pig prices and margins

Pig prices were assessed for each type of production from market reports of the MLC and farming press related to prices from the scheme. The average net prices received by the herds in the scheme for the latest recording year were deducted from market report quotations (usually gross prices) averaged over the same year. The difference between the two allows for haulage and marketing charges and these amounts were used to adjust current quotations to provide up to date pig prices on the same basis as those recorded in the scheme. The adjusted pig prices applied to both spot and spread costs and give updated margins per pig as illustrated in Table 15.4.

The examples show that production costs for March 1992 had increased above the levels for the year ended 30 September 1991, used for the data base, partly through rising costs and partly through seasonal adjustment for the winter period. Estimated costs of producing 30 kg weaners were £1.52 higher for spot and £1.70 higher for the spread method. The latter contained several months with the 103 per cent seasonal adjustment, while spot costs applying to the month of March were at 101 per cent. For breeding and feeding for cutter production spot costs were £3.10 per pig higher but spread costs, which included a lower weaner cost from four months previous, only £2.67 higher. Most noticeable was the difference in costs of feeding purchased weaners for cutters. A substantial rise in weaner prices, reflecting higher finished pig prices, was largely responsible for the £9.24 increase in spot costs, while spread costs included a much lower cost weaner from three months previous, which almost offset the increase in production costs.

Despite higher costs, pig prices had risen even more, and margins for March were much improved from the dismal average level of 1991. Spot margins for feeding-only herds producing cutters had increased at a lower rate as weaners were now far more expensive. The spread margin, based on a much lower weaner cost from three months earlier, was the best of all. This clearly demonstrates the importance of distinguishing between spot and spread costs. For a feeding-only unit producing cutters the margin for March at current prices (spot) was £4.31 per pig but the margin on costs incurred by the pigs just slaughtered (spread) was £13.35.

Table 15.4 A specimen of estimated average profitability of
pig production from updated costs and returns
(seasonally adjusted)

	March 1992 Spot[a]	March 1992 Spread[b]
Breeding - weaner sales		
Weaners (30 kg)	£	£
Feed	19.28	19.32
Labour	6.77	6.85
Other costs	6.55	6.61
Stock depreciation	1.29	1.29
Total costs	33.89	34.07
Net return	40.60	40.60
Margin per pig	6.71	6.53
Breeding and feeding		
Weaner (8 wks - 18 kg)	£	£
Feed	14.76	14.77
Labour	5.94	6.01
Other costs	6.16	6.21
Stock depreciation	1.02	1.02
Weaner costs c/f	27.88	28.01
Finishing - cutters (66 kg)	£	£
Weaner - home-bred b/f	27.88	27.20[c]
Feed	29.48	29.66
Labour	3.79	3.83
Other costs	4.38	4.43
Mortality charge	.70	.68
Total costs	66.23	65.80
Net return	78.64	78.64
Margin per pig	12.41	12.84
Feeding-only - cutters (66 kg)	£	£
Weaner - purchased (30 kg)	41.60	32.47[d]
Feed	24.90	25.09
Labour	3.19	3.24
Other costs	3.89	3.93
Mortality charge	.75	.56
Total costs	74.33	65.29
Net return	78.64	78.64
Margin per pig	4.31	13.35

(a) Costs at prices for current month
(b) Costs spread over the production cycle
(c) Home-bred weaner costs from four months previous
(d) Purchased weaner cost from three months previous

At the end of the year a comparison can be undertaken to measure the accuracy of the updated monthly estimates against the actual results achieved by the herds in the scheme for the corresponding year. The checks for 1991 are shown in Table 15.5.

Table 15.5 A comparison of a year's monthly updated estimates (spot) with the actual annual results from the scheme 1991[*]

Breeding and feeding (Feeding-only in parentheses)

	Total costs		Net returns	
	Update	Actual	Update	Actual
Weaners (30 kg) per pig	£32.72	£32.37	£31.86	£32.68
	p	p	p	p
Porkers per kg dwt	103.9	102.4	101.3	101.0
	(106.8)	(105.8)		
Cutters per kg dwt	95.0	95.7	99.1	97.9
	(97.5)	(98.7)		
Baconers per kg dwt	94.9	94.2	99.4	98.4
	(97.7)	(97.1)		

*Year ended 30 September

In all categories the updated estimates averaged over the year were quite close to actual results achieved. As the updating was based on the physical performance of the previous year and adjusted according to price changes monitored each month, it would be surprising if the updates for the year were exactly the same as those from the recorded herds. Performance standards do change from time to time, long-term improvements have been substantial, but nowadays changes between years are usually marginal. Furthermore, some farmers benefit over spot feed prices from forward buying and it may be advisable to take this into account should prices change drastically, as those of soya bean meal have occasionally done. Similar comparisons in previous years have nearly always shown the updated estimates to be within a few pence per kg deadweight of the actual results produced for the year. This method of updating has proved a fairly accurate assessment of pig profitability, and it can be undertaken promptly each month rather than having to wait until the recording year has ended.

Alternative method

Updating costs and returns quickly indicates the current profitability of pig production, information which can be especially useful in times of rapid change. Such up to date information is needed to assist in making decisions concerning present and future policies both nationally and for individual units. However, it is a rather complex and laborious task to undertake and few pig producers are likely to have sufficient time to embark on their own project, even if all of the relevant data were readily available.

As an alternative, a simple yardstick to assess current production costs can be useful, especially when feed prices fluctuate. The average quantity of feed required will reflect the standard of efficiency achieved and should be available from results of herd recording. As most finished pigs are paid for by deadweight it is helpful to know how much feed was used to produce 1 kg deadweight of pigmeat. Simply divide the total quantity of feed used per pig for breeding and finishing by its deadweight. Recording results usually give a feed conversion rate by liveweight but the deadweight figure can be calculated by multiplying the conversion rate by the liveweight gain of feeder pigs, adding the quantity required for the weaner (breeder-feeder herds) and dividing the total feed by the pig's deadweight.

For breeding and feeding herds the total quantity of feed required per kg deadweight (Table 15.6) was on average virtually the same for all types of production for the five years 1987-91 in the Cambridge scheme. The worsening conversion rates for the heavier pigs were offset by spreading the quantity of feed per weaner over larger carcases, which benefited from higher killing-out percentages.

Although there was slight variation between types of production and between years, the five-year 1987-91 averages were the same at 4.1 kg of feed per kg deadweight. From 1990-91 a figure of 4.0 kg seems more appropriate.

Table 15.6 Total feed required per kg deadweight
(breeding and feeding)

	Porkers kg	Cutters kg	Baconers kg
1987	4.1	4.2	4.1
1988	4.2	4.2	4.2
1989	4.1	4.1	4.1
1990	4.0	4.0	4.1
1991	4.1	4.0	4.0

The next stage is to assess the current cost of feed per tonne, or per kg, of all feed being used. An average for breeder-feeders herds of say £160 per tonne (more for porkers) is equal to 16p per kg and to calculate feed costs per kg deadweight one need only multiply this figure by the quantity used. For this example, using 4 kg, the cost of feed would be (16p x 4 kg) 64p for breeding and feeding cutters or baconers. Porkers would be nearer to (17p x 4 kg) 68p per kg deadweight. Then labour and other costs, together with an allowance for stock depreciation and feeder mortality, need to be added. In 1991, they averaged about 32p per kg deadweight (35p for porkers) to give total costs of (64p + 32p) 96p per kg for cutters and baconers and (68p + 35p) 103p for porkers. A seasonal adjustment can be applied to these costs if required.

When these costs are related to the net price per kg deadweight received for pigs sold, the margin will be revealed. If, for example, the price was 110p (porkers 113p) margins would be 14p per kg for cutters and baconers (10p for porkers). The amount per pig can be calculated by multiplying the margin per kg by the deadweight per pig.

Similar brief assessments can be undertaken to cover those starting with a purchased weaner and feeding-only. The quantity of feed required for the feeding-only stage (Table 15.7) will be less than that for the breeding and feeding herds and the cost of feed per tonne is also usually slightly less.

Table 15.7 Feed required per kg deadweight
(Feeding-only from 30 kg weaners)

	Porkers kg	Cutters kg	Baconers kg
1987	2.2	2.5	2.6
1988	2.3	2.6	2.7
1989	2.2	2.5	2.6
1990	2.1	2.5	2.6
1991	2.2	2.4	2.5

Assuming an average for feeding-only herds of £150 per tonne (15p per kg), the cost of feed per kg deadweight for feeding-only herds producing cutters or baconers would be (15p x say 2.5 kg) 37.5p. The cost for porkers might be £160 per tonne (16p x 2.2 kg) 35.2p per kg deadweight. Labour, other costs and mortality would average about 12p per kg deadweight for porkers and 13p for cutters and baconers, to give feeding costs of about 47p for porkers and 50p for cutters and baconers. These costs for feeding-only herds were considerably less than for breeding and feeding but here the cost of the purchased weaner must be added. This is simply the cost per weaner divided by the deadweight of pigs produced and, therefore, the cost per kg falls as the weight increases. For example, a £35 weaner equals 63p per kg for a 55 kg deadweight porker, 53p for a 66 kg cutter and 51p for a 69 kg baconer. In this case total costs per kg deadweight amount to 110p for porkers, 103p for cutters and 101p for baconers. Net returns must exceed these costs to produce a surplus margin.

These examples of simple yardsticks to indicate current profitability are based on average performance. Individual results will vary considerably but pig producers should be able to obtain quickly an estimate of their own production costs and margins by this method if they know the standard of performance achieved by their herd. For breeding and feeding herds, the quantity of feed required per kg of finished pig must cover the amount used to produce the weaner and for rearing the weaner to slaughter weight. The range from herd to herd is quite large as shown in Table 15.8, giving the total quantity of feed required per kg deadweight from varying levels of the amount per weaner and conversion rates. This example is based on breeding and feeding for cutter production.

Table 15.8 Total feed required per kg deadweight for cutters (66 kg) at varying quantities per weaner and conversion rates (breeding and feeding)

Conversion rate	kg per weaner (8 weeks - 18 kg)				
(per kg lwt gain)	60	70	80	90	100
	Quantity per kg deadweight				
	kg	kg	kg	kg	kg
2.2	3.24	3.39	3.55	3.70	3.85
2.4	3.45	3.61	3.76	3.91	4.06
2.6	3.67	3.82	3.97	4.12	4.27
2.8	3.88	4.03	4.18	4.33	4.48
3.0	4.09	4.24	4.39	4.55	4.70

A feed requirement of 90 kg per weaner, coupled with a 2.8 conversion rate gives a total of 4.33 kg of feed per kg deadweight. A better performance of 70 kg per weaner with a 2.4 conversion rate gives a much lower total 3.61 kg per kg deadweight. At 16p per kg (£160 per tonne), this difference is worth 10.8p per kg deadweight, or £7.13 per 66 kg deadweight pig.

Table 15.9 shows the quantity of feed per kg deadweight of finished pig for feeding-only herds from varying conversion rates. It is assumed that weaners were bought in at 30 kg.

Feed requirements for a feeding-only herd producing 66 kg deadweight cutters with a conversion rate of 2.8 is equal to 2.46 kg per deadweight of finished pig. If the feed cost 15p per kg (£150 per tonne) the cost per kg deadweight for feed comes to 37p. Labour, other costs and mortality charge say 13p per kg and the weaner overhead (say £35 spread over 66 kg) equals 53p per kg. Total production costs, therefore, amount to (37p + 13p + 53p) 103p per kg and the net price per kg deadweight received must exceed this sum to leave a profit margin.

Table 15.9 Quantity of feed required per kg deadweight from
varying conversion rates
(feeding-only herds - from 30 kg weaners)

Conversion rate (per kg lwt gain)	Porkers (55 kg dwt)	Cutters (66 kg dwt)	Baconers (69 kg dwt)
	kg	kg	kg
2.2	1.80	1.93	1.97
2.4	1.96	2.11	2.16
2.6	2.13	2.28	2.34
2.8	2.29	2.46	2.52
3.0	2.45	2.64	2.70

For breeding herds selling weaners a quick assessment of costs is also possible, though there could be more variation depending on size of weaners and the proportions of comparatively expensive piglet feeds to sow feed used. The necessary calculation involves valuing the average quantity of feed used to produce a weaner to sale weight, at current prices, and adding an estimated sum to cover labour, other costs and stock depreciation. Average requirements of sow and piglet feed per 30 kg weaner are given in Table 15.10.

Table 15.10 Average quantity of sow and piglet feed
used per 30 kg weaner

	kg
1987	103
1988	102
1989	105
1990	103
1991	103

Because probably two types of piglet feed and at least one feed for sows would be used, the calculation of a weighted average feed cost per tonne may be rather complex. Assuming an average cost of £185 per tonne (18.5p per kg), 103 kg would be worth about £19 per 30 kg weaner. Average labour, other costs and stock depreciation amount to some £15 per pig; outdoor units about £1 less. Total costs come to approximately (£19 + £15) £34 per 30 kg weaner, or £33 for outdoor units, and the sale price must exceed this amount to leave a profit margin.

In times of changing feed and pig prices, these quick calculations give an indication of current margins. The examples included here are only a guide, and greater accuracy should result if individual producers recorded their herds and had knowledge of recent performance standards achieved to use instead of standards. Violent price fluctuations can be very worrying but an updating to measure current profitability, by whatever means, should at least ensure that management decisions are made from a sound basis and not from ill-informed opinions.

Chapter 16

A Method of Calculating Weaner Prices

The escalation of costs and returns in pig production in the early 1970s caused considerable problems for those wishing to trade in weaners. Auction markets, once the generally accepted barometer for measuring free trade, were already in decline and vulnerable to manipulation by a few large dealers. The quality of weaners was variable and supplies irregular. Speculation was common and many breeders considered they suffered badly from not always receiving a fair proportion of the finished pig price. Several producer groups were created for marketing livestock and most formed a committee to organise their weaner trade. The committee usually consisted of equal numbers of breeders, who were selling, and feeders, who were buying, with group management providing the secretarial cover. The work functioned smoothly while prices remained stable but when costs and prices were rising difficulties became apparent. Some thought price changes were excessive and others complained that there had been no constant relationship between the weaner price and the finished pig price. The meetings were often harassing and time consuming, with few completely satisfied by the outcome. Long-established relationships were strained through arguments over weaner prices.

Against this background and following pressure from members of the groups, efforts were made then to find methods of calculating weaner prices that were fair for both breeders and feeders. The Cambridge scheme showed that fluctuations in profitability caused by changes in costs and pig prices applied to all producers. For those that undertake only one stage of production, either breeding weaners for sale or feeding-only, there is an additional hazard - the price of weaners. The profitability of selling weaners compared with finishing them is usually related to weaner prices. When weaners are cheap, feeders have the advantage, and when they are dear, the breeder benefits. There have been times when weaner prices were so high that the disparity in profits between breeders and feeders was substantial, with the breeders doing well while the feeders were losing money. At other times when the demand for weaners sagged and prices were low, the positions were reversed and the feeders came off best.

Equitable shares

For many years an accepted and fair method of comparing profitability between herds of different sizes and systems has been in relation to the value of output. There seemed no reason why weaner prices should not be calculated on the same basis. Weaner prices based on the overall costs and returns can avoid some of the variations in profitability between breeders and feeders to secure margins more like those of the breeder-feeders, who are insulated from weaner prices. This could be achieved by adopting a method of assessing weaner values which equitably shares any profit to give both parties the same margin per £100 output.

A method which provides the same margin per £100 output for the breeder and the feeder should include current costs of production, as well as prevailing returns. In the method operated at Cambridge for several organisations, average results from the scheme for the latest year available were updated each month to allow for recent changes in feed, labour and other costs to provide current costs. Actual costs do, of course, vary considerably from herd to herd and the use of average figures allows the producer with below average costs to benefit from his achievements and is an incentive to encourage efficiency.

To maintain harmony between breeders and feeders the weaner prices must be realistic and in the long-term reasonably similar to weaner prices reported in the farming press. Free market prices are often influenced by supply and demand which can occur irrespective of prevailing costs and returns. Many buyers and sellers of weaners are prepared nowadays to disregard short-term differences for the sake of long-term stability and permanent arrangements. However, loyalty can be severely tested, even on a contractual basis, when the breeder could sell his weaners at a higher price to a different buyer, or the feeder could buy them cheaper. In such cases it may be necessary to include a stabilising adjustment to limit the amount of any difference.

The relationship for this method of calculating weaner prices may be expressed by the equation shown below.

$$Y = \left((X_1 \times X_2) - (X_3 + X_4) \times \frac{X_5}{X_4} \right) + X_5 = K$$

For the stabiliser $\quad Y = K \pm X_6$

K   = Calculated weaner price

Y   = Weaner price (over weight range, say 25-35 kg)

X_1 = Average slaughter weight of pigs produced

X_2 = Price per kg deadweight

X_3 = Marketing charges

X_4 = Total costs of breeding and feeding per pig

X_5 = Breeding costs per weaner (over weight range, say 25-35 kg)

X_6 = Stabilising arrangement

Data base

Average results from the scheme for the latest year available, were used to calculate the costs of producing 30 kg weaners for the breeders and for finishing 30 kg pigs to 89 kg liveweight for the feeders. The latter liveweight related closely to the average of pigs produced by the feeder herds in the scheme, and the deadweight equivalent (66 kg) was the estimated average of all pigs produced (AAPP) used to value the finished pig.

The 1990 results

The average results from the scheme for the recording year ended 30 September 1990, adjusted for minor differences in pig weights, gave the following costs per pig.

	Weaners		Feeders
Liveweight at sale	30 kg	Liveweight at start	30 kg
		Liveweight at finish	89 kg

Costs per weaner	£	Costs pig	£
Feed	18.01	Feed	24.36
Labour	6.12	Labour	2.85
Other costs	6.17	Other costs	4.10
Stock depreciation	.76	Morality charge	.77
Total breeding	31.06	Total feeding	32.08

Feed used

The main types of feed used by the herds in each group in achieving the average performance in 1990 were as follows.

	Breeders %	Feeders %
Sow compounds	46.2	-
Piglet compounds	33.5	-
Feeder compounds	-	36.8
Concentrate mixes	2.8	4.2
Wheat	9.5	25.7
Barley	3.8	15.0
Milling offals	2.8	4.9
Soya bean	1.2	11.8
Fish meal	.2	1.6
	100	100
Average cost of all feed used per tonne	£174.14	£143.89

Updating costs

The annual costs were updated each month throughout the following year for changes in the price of feed, labour and other costs. For feed, the latest quotations averaged over the month in question were related to the average of the same quotes over the last recording year and the difference used to update last year's costs. Labour costs were updated from awards announced by the Agricultural Wages Board and from any changes in the employer's share of National Insurance contributions. Allowances were made for an increase in other costs. An example of updating feed costs per tonne for breeders and feeders for the month of May (1991) is given in Table 16.1.

Table 16.1 An example of updating feed costs
for weaner price calculations

	Average quotations for previous year (1990)	Quotes for May 1991	Change per tonne	Value per tonne[a]	
				Breeders	Feeders
	£	£	£	£	£
Breeder compounds	153.44	168.60	+ 15.16	+ 7.00	-
Piglet compounds	269.37	283.19	+ 13.82	+ 4.63	-
Feeder compounds	170.89	184.74	+ 13.85	-	+ 5.10
Breeder conc mixes	280.81	276.18	-4.63	-0.13	-
Feeder conc mixes	290.57	285.02	-5.55	-	-0.23
Wheat	109.77	128.82	+ 19.05	+ 1.81	+ 4.90
Barley	101.69	114.56	+ 12.87	+ 0.49	+ 1.93
Milling offals	104.80	123.85	+ 19.05	+ 0.53	+ 0.93
Soya bean	154.47	150.40	-4.07	-0.05	-0.48
Fish meal	360.32	360.00	-0.32	0.00	- 0.01
				+ 14.28	+ 12.14
Average cost for year 1990				174.14	143.89
Updated cost for May 1991				188.42	156.03

(a) At percentages of types used in 1990.

Compared with the recording year of 1990 feed costs had risen by £14.28 per tonne for the breeders and £12.14 for the feeders by May 1991. For the same period labour costs had increased by 7.7 per cent (a further increase became effective in June) and other costs were estimated at 14p per pig higher for both the weaner and the feeder. The breeding stock depreciation was assumed to be unchanged, while feeder mortality was 18p per pig lower as weaner prices had fallen.

Having updated feed costs per tonne, the next stage translated the increase into costs per pig at the latest known feed requirement rates. Some adjustments were necessary to the previous year's average results to match the 30 kg required for the calculated weaner price. Feed requirements for updating costs were as follows.

For breeding	kg
Sow feed per weaner produced	61.5
Piglet feed per weaner	41.9
Total feed per 30 kg weaner	103.4

For feeding	kg
Total quantity per 30 to 89 kg pig	169.4

The 1990 average costs had increased by the following amounts by May 1991.

Breeding	£
Feeding £14.28 x 103.4 kg =	1.48
Labour	.47
Other costs	.14
	2.09

Feeding	£
Feed £12.14 x 169.4 kg =	2.06
Labour	.22
Other costs	.14
	2.42

Seasonal variation

So far this has been an updating of last year's annual results. Performance and production costs vary however throughout the year according to season. More pigs are reared per sow in summer than in winter and less feed is required for both breeding and feeding stages of production. In order to provide realistic assessments of costs and margins each month, the following seasonal adjustments were made to the annual and updated costs to take account of this variation.

Month	Jan	Feb	Mar	Apr	May	June	Jul	Aug	Sep	Oct	Nov	Dec
% of	%	%	%	%	%	%	%	%	%	%	%	%
costs	103	103	101	99	97	97	97	97	99	101	103	103

Production cycles

The monthly updated and seasonally adjusted costs were then spread over the expected production cycle. The percentages of costs applicable to each month were assessed as follows.

Breeding Month	%		Feeding Month	%
1	10		1	21
2	10		2	33
3	10		3	46
4	20			100
5	23			
6	27			
	100			

Marketing charges

Representative marketing charges incurred separately by the breeder and the feeder to cover weaner and finished pig haulage, insurance, levies, etc., were added to costs at a rate appropriate to both parties. These marketing charges were reviewed periodically and changes introduced when necessary. In the example given in Table 16.2 it was assumed that the feeder paid all charges including haulage of weaners.

Table 16.2 Example of calculated weaner price for June 1991

	Breeding £	Feeding £
Last year's average costs[a]		
Breeding costs per weaner (30 kg)	31.08	-
Feeding costs per pig (30-89 kg lwt)	-	31.60
Adjustments for recent cost changes	1.67	1.80
Updated costs of production[a]	32.75	33.40
Haulage and marketing charges	-	2.86
Total costs[b]	32.75	36.26
Percentage of total net costs[c]	49.4 %	50.6 %

	£
Value per pig (66 kg dwt @ 117.82p)	77.76
Total costs - breeding and feeding	69.01
Margin per pig	8.75

	Breeding	Feeding
Updated costs of 30 kg weaner	32.75	-
Share of margin (49.4 and 50.6%)	4.32	4.43
Calculated price per 30 kg weaner	37.07	-

Marginal price per kg for weaners of different weights   58p

Batch pricing : £19.67 per pig plus 58p per kg for the total weight.

(a)   Seasonally adjusted and covering production cycle
(b)   Excluding interest charges
(c)   Total costs less stock depreciation, mortality charge, haulage and marketing charges

Finished pig value

The finished pig was valued according to the average all pigs price (AAPP) published by the MLC and represented the weighted average price for all types of finished pigs (other than those used for breeding) sold in the UK each week. To smooth any fluctuations in price, the AAPP used was the average of the latest four weeks available.

Weaner price statement

The weaner price calculations were usually undertaken at the end of the month prior to that to which they applied and a statement produced by the operative date. Details given related to 30 kg weaners but a marginal rate per kg was also shown for adjusting the price for pigs sold at weights above or below 30 kg. The use of this marginal price per kg ensures that margins per £100 output remain the same for both breeder and feeder, whatever the weight of weaner. Calculated weaner prices fluctuate, due to changes in costs and finished pig prices, as does the marginal rate per kg. An alternative method of payment, frequently used for batch pricing, is also given. This consists of a basic price per pig, to which the marginal rate per kg is added for the total weight of weaners in the batch. Details of a calculated weaner price would be something similar to the example in Table 16.2.

Costs for the breeder and the feeder are added together and deducted from the value of the finished pig. The resulting margin per pig is shared between the two parties on the basis of the contribution each makes to total net production costs. The breeder's share of the margin is added to the updated costs of producing the weaner, to give the calculated price of the weaner. Should the value of the finished pig be less than total costs, the loss is shared on the same percentage basis and the breeder's share then deducted from the weaner costs. The weaner price in this example was £37.07. The margin, shared £4.32 to the breeder and £4.43 to the feeder, provides both parties with the same return of £11.90 per £100 of output (Table 16.3).

This seems a fair method of assessing weaner prices and sharing what money there is in pigs between breeders and feeders. Although market prices will vary according to supply and demand, usually with little regard for production costs, those looking for a stable relationship should find this method attractive. Only in exceptional circumstances, such as proven superior quality of weaners or trading in large numbers, is there any justification for one party to have a larger margin per £100 output than the other.

Table 16.3 Margins per pig and per £100 output

	Breeding 30 kg	Feeding 30-89 kg
Per pig	£	£
Value at sale	37.07	77.76
Weaner cost	-	37.07
Haulage and marketing expenses	-	2.86
Stock depreciation	.76	-
Mortality charge	-	.59
Output	36.31	37.24
Costs (net)	31.99(a)	32.81(b)
Margin	4.32	4.43
Per £100 output	£	£
Costs (net)	88.10	88.10
Margin	11.90	11.90
	100	100

(a) Updated costs £32.75 less stock depreciation
(b) Updated costs £33.40 less mortality charge

These calculations relate to a weaner of 30 kg and, of course, not all weaners change hands at precisely this weight. Some will be lighter and others heavier. The lighter weaners should cost less to produce, while the heavier ones cost more. To calculate the appropriate weaner price for pigs of different weights, the method is the same as before. Because of different costs involved the percentage share of total net costs for the breeder and the feeder will also be different. Total costs (breeding and feeding together) should be the same throughout but with lighter weaners the breeder's costs and the percentage share of the total are less. On the other hand, the feeder starting with a smaller weaner will incur higher costs in reaching slaughter weight and, therefore, receives a correspondingly greater share of the percentage of total costs. Prices for weaners of varying weights are shown in Table 16.4.

Table 16.4 Calculated price for weaners of varying weights

Weaner weight	Weaner[a] price	Production costs Breeding	Production costs Feeding	Breeding costs as % of total	Share of margin Breeding	Share of margin Feeding	Margin[b] per £100 output
kg	£	£	£	%	£	£	£
26	34.75	30.71	38.30	46.2	4.04	4.71	11.90
28	35.91	31.73	37.28	47.8	4.18	4.57	11.90
30	37.07	32.75	36.26	49.4	4.32	4.43	11.90
32	38.23	33.77	35.24	50.9	4.46	4.29	11.90
34	39.39	34.79	34.22	52.5	4.60	4.15	11.90

(a) The weaner price equals breeding costs plus the breeding share of the margin. When pig margins are in deficit the breeder's share is then deducted from breeding costs.

(b) The same for breeder and feeder

Throughout this weight range the margin per £100 output is the same for both the breeder and the feeder at £11.90. If weaner prices elsewhere were trading at less than those given here, it means that, on average, the breeders selling these weaners were at the time receiving a smaller share of the margin per £100 output than the feeders.

During the four years 1988-91 the margin per pig to be shared ranged from a surplus of £25 to a deficit of £13. Continuous monitoring of prices to update costs ensures that the percentage share of costs to the breeder and the feeder for allocating the margin are fair to both sides. These percentages are not fixed amounts but frequently change as costs rise or fall. For instance, if compound feeds rise in price more than straight feeds, then breeding costs, with a higher percentage of compounds used, will increase more than feeding costs and, therefore, form a greater proportion of the total.

Weaner prices compared

In the long term, the calculated weaner prices, providing equal margins on output for both breeders and feeders, are likely to be similar to those paid by the major buyers but occasionally differences do occur. When the demand for weaners exceeds the supply, market prices rise, regardless of the level of production costs and returns for finished pigs. In reality, the feeders who need weaners are competing for a scarce commodity, causing

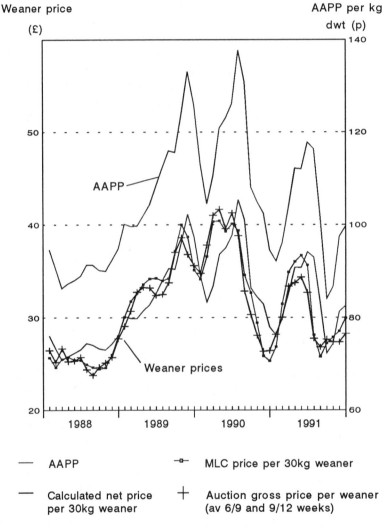

Figure 16.1 A comparison of weaner prices

prices to rise. If they decide to buy at the higher price, it means they forfeit some of the profit margin which should eventually have been theirs.

The opposite situation arises when positions are reversed. An over-supply of weaners, or a fall in demand, often means that the breeder must take lower prices in order to dispose of the weaners under free marketing. If this is below the calculated price, the breeder then foregoes some of the profit. Most breeders have limited accommodation for their pigs and, once they have reached sale weight, can seldom retain them for more than a few days hoping for a better price.

A comparison of calculated weaner prices with auction markets and MLC price reports elsewhere is given in Figure 16.1.

The calculated weaner price follows the movements of AAPP more closely than other prices, which was to be expected. Over the four years there have been periods when the calculated price was higher than others and some when it was lower.

Stabilising adjustment

When calculated weaner prices differ from those elsewhere for prolonged periods, such as occurred during 1990, then a stabilising adjustment may be necessary to retain the loyalty of individuals to the trading agreement. Any such adjustment should be agreed in advance and is best left to the discretion of the management to operate. It needs to be based on circumstances within the organisation concerning the supply and disposal of the weaners available and relative prices locally. A stabiliser could function on the basis of limiting any difference between the calculated price and a representative price elsewhere (say auction market or MLC reported price). Perhaps a variation exceeding 10 per cent above or below the calculated price should trigger a stabilising adjustment to be applied at a rate of half the difference between the two. An adjustment does mean, of course, that one party then makes more profit at the expense of the other.

Pig producers who breed weaners for sale, or who buy them for finishing, have a choice over the way future weaner prices are fixed. They can co-operate with each other and arrange for weaner prices to be assessed on a basis which gives both breeder and feeder an equal return in relation to output, or they can take the going rate on offer for weaners, which is influenced by supply and demand, often irrespective of prevailing costs and finished pig prices. Those that adopt the former choice will at least have the satisfaction of knowing that weaner prices are fair to both parties concerned.

Alternative yardstick

With the demise of the Cambridge scheme the data for calculating fair weaner prices may no longer be as readily available, though it should be possible for other pig surveys to produce something similar. The essential components for a satisfactory undertaking of this project are:

1. A reliable detailed data base of representative standards of performance and production costs for:
    a. breeding herds selling weaners
    b. feeding herds buying weaners for finishing

2. Records for monitoring changes in prices to update production costs and determine current finished pig prices

An alternative can be derived from the results of past calculations which established a relationship between weaner prices and finished pig prices. No short-cut method will be as good as the detailed calculations and could be subject to some variation from time to time. Results from previously calculated weaner prices give the base rate : AAPP ratio and the marginal rate : AAPP ratio over five years as shown in Table 16.5.

In 1988 the base rate per pig for calculated weaner prices averaged 14.93 times the AAPP and increased steadily each year to 17.07 times in 1992. This rise in the ratio was due to the changing structure of costs in pig production as labour and other costs increased, while feed costs remained relatively stable, to form a greater proportion of total costs. If all costs stayed the same, or increased at the same rate, then the ratio would remain constant. Labour and other costs are greater for breeding stock (41 per cent of the total in 1991) than for feeders (24 per cent), so the recent increases have raised costs more for breeding than for feeding; hence the rising ratio. Although the rise has been more in some years than others, it amounts to an average 0.5 increase per year over the five-year period. For this method to work satisfactorily in future it will be necessary to establish each year the rate of any change to the structure of costs and how this affects breeding costs as opposed to feeding costs. If, for example, feed costs fall while labour and other costs continue to rise, the ratio will also rise, but if feed should increase more than labour and other costs, then it will decline.

Table 16.5 Ratio of base rate and marginal rate
to AAPP for calculated weaner prices

	Annual average	Monthly range Maximum	Minimum
1988 *			
Base rate : AAPP ratio	14.93	15.33	14.67
Marginal rate : AAPP ratio	0.487	0.496	0.481
1989			
Base rate : AAPP ratio	15.70	16.23	15.41
Marginal rate : AAPP ratio	0.488	0.492	0.481
1990			
Base rate : AAPP ratio	16.06	16.42	15.57
Marginal rate : AAPP ratio	0.486	0.496	0.476
1991			
Base rate : AAPP ratio	16.75	17.15	16.20
Marginal rate : AAPP ratio	0.485	0.493	0.473
1992			
Base rate : AAPP ratio	17.07	17.38	16.50
Marginal rate : AAPP ratio	0.486	0.490	0.478

*Jan-Dec year.

The weaner price base rate : AAPP ratio also varied month by month as shown by the range in Table 16.5. Usually the ratio increased as the year progressed, though not in the last two months of 1992 when devaluation of the green pound raised feed costs substantially. For those wishing to assess weaner prices on this basis, it is suggested that a ratio should be agreed at the beginning of each year, from knowledge of the latest cost structure, and left to operate for the remainder of the year. In 1992, for example, an appropriate ratio to give a monthly weaner price base rate would have been 17 times the latest AAPP (or average of the last four weeks' AAPP).

The marginal rate per kg, to apply to the total weight of weaners, was far more consistent over the five-year period examined. The annual average stayed at just half of the AAPP rate throughout and monthly variations were quite small. If nothing else, this study clearly demonstrates that a weaner price marginal rate of half the AAPP is far more accurate and better than the 50p per kg most buyers use continually because of ease of administration. When weaner prices fluctuate between £26 and £42 a head, as they did in 1990, then a marginal rate of 50p per kg in both cases cannot be correct. This standard 50p rate favours the breeder when prices are low and the feeder when they are high. On occasions this can lead to disagreements when one party prefers to trade lightweight weaners while the other wants them heavier. A marginal rate per kg of AAPP times 0.5 would avoid such difficulties and provide a more realistic reflection of pig prices.

An alternative method of pricing weaners in 1992 based on AAPP x 17 as the base price and AAPP x 0.5 as the marginal price per kg would have been as follows:

(AAPP x 17) + (AAPP x 0.5 x weaner weight)

If AAPP was 112p, a 30 kg weaner would be valued at £35.84 (112p x 17 plus 112p x 0.5 x 30). Other examples are given in Table 16.6.

The AAPP in 1992 averaged 115.24p and on this method would have produced weaner prices which were on average £36.88 for a 30 kg pig. Prices in the Eastern region reported by the MLC for the year averaged £36.55 but this similarity could have been just coincidence. Although the marginal rate assessment works well, the difficulty of deciding the right multiplier for calculating the base price may restrict the use of this method in future, though similar versions are already in operation by some producers. For good trading relations it is essential to agree a number each year and a review should be undertaken periodically for any changes in feed, labour and other costs that may affect the overall structure of costs.

Table 16.6 A method of relating weaner prices
to finished pig prices (AAPP)
(examples for 30 kg weaners)

AAPP	Base[a] price	Total[b] marginal price	Weaner price (30 kg)
p	£	£	£
80	13.60	12.00	25.60
90	15.30	13.50	28.80
100	17.00	15.00	32.00
110	18.70	16.50	35.20
120	20.40	18.00	38.40
130	22.10	19.50	41.60

(a) AAPP x 17
(b) AAPP x 0.5 x 30

Perhaps a safer method of arriving at weaner prices would be to follow the going rate as published in the farming press, though this means abandoning profit sharing ambitions. Prices are usually quoted for 30 kg weaners and by using half of the current AAPP for the marginal rate, the price for lighter or heavier weaners can be quickly calculated. For example, if the quoted reference price for 30 kg weaners was £35 and the AAPP was 110p, a 28 kg weaner would be valued at £35 less £1.10p (0.5 x 110p x 2) = £33.90. The same adjustment of half AAPP times 30 would give a base price per weaner for those preferring a batch pricing method of a basic rate per weaner plus half AAPP for the total weight of weaners.

Over the years many producers involved in selling or buying weaners have expressed a preference for prices to be fixed on a profit sharing basis. Free market prices often give one party an advantage at the expense of the other. Most producers would prefer a reasonable share of what money there is in pigs all the time to a large share sometimes and a small share or nothing at all at other times. Perhaps one of the several organisations now monitoring the progress and performance of pig production could undertake a fair weaner pricing service for producers.

Chapter 17

Future Prospects

Pig production in the run-up to the year 2000 will be very different from that which existed in earlier decades during the lifetime of the Cambridge scheme. There have been many changes in production methods and, in recent years, restrictions on management practices and doubtless there are more to come. These involve legislation on planning, environment, welfare and safety at work. All are no doubt desirable but most are costly to comply with, especially the enforced controls on pollution. The quality and safety of our food, however, are paramount and high standards should enable British pigmeat to establish a favourable identity to compete against other EC nations.

Changing structure

In the 1950s some 150,000 farmers in England and Wales kept about four million pigs between them. The pigs were accommodated in a wide variety of housing but mainly in whatever farm buildings were available. Most herds were small; nearly 140,000 (93 per cent) were under 100 pigs in total while only 77 (0.05 per cent) had more than 1,000 pigs. Since then the number of pig producers has fallen drastically and by 1992 was down to just over 13,000. Pig numbers increased to 7.3 million by 1973 but within a couple of years had fallen back to 6.5 million and have since remained at around this level. Units are now much larger and most are housed in buildings specially designed for the purpose. In 1992 there were still 7,500 small herds (57 per cent of the total) with less than 100 pigs but the larger units of over 1,000 pigs had increased to 1,900 (14 per cent). The small units, accounting for half of the pig producers, had only 2 per cent of the pigs in England and Wales, while the largest 14 per cent of units had 73 per cent of the pigs.

This trend of concentration into fewer hands is likely to

continue and by the end of the millennium the number of pig producers in England and Wales could be down to 11,000. The number of pigs is likely to remain at about 6.5 million. Pollution controls and the ban on the use of sow stalls could inhibit existing large herds from expanding for the time being, but their number could increase marginally to around 2,000 (18 per cent of the total) and they would probably then contain about 5 million (77 per cent) of the country's pigs. Because of the uncertainty of future cereal production some arable farmers may be tempted to start pig farming as a means of increasing the value of their cereals. Otherwise new entrants to pig production are expected to be few. Many small units of less than 100 pigs have already disappeared and this exodus is likely to continue as more farmers become disenchanted with looking after a few pigs seven days a week often for little reward. By the year 2000 their number could fall to 5,700 (52 per cent) with 110,000 pigs (1.7 per cent). This leaves the middle size group, those with more than 100 pigs but less than 1,000 pigs. Their number is also likely to decline to a forecast 3,300 (30 per cent) with a total of about 1.4 million pigs (21.3 per cent); see Table 17.1.

By the year 2000 the small units of less than 100 pigs are still expected to be the largest group, with just over half of the total, but their number since 1970 is likely to have fallen by 39,000, or 87 per cent. In 1970 they had between them nearly one-fifth of all pigs in England and Wales; in 2000 they could have less than 2 per cent. The middle size group should remain the second largest group with about 30 per cent of units, though their number could be some 11,000 below the 1970 level, a fall of 77 per cent. This group in 1970 had just over 60 per cent of the pigs, but their share has fallen steadily since and by the year 2000 could be down to only a little over 20 per cent of the total.

The large herds, those of 1,000 pigs and over, are likely to increase in number during the 1990s at a similar rate to that of the previous decade, mainly from expansion of herds in the next size group and further involvement in production by feed companies and pig procurers in establishing more new units. This group is the only one to have increased in number of units for many years; they already dominate home pig production and are likely to consolidate this position to hold over three-quarters of the nation's pigs.

The total number of pigs is not expected to change much by 2000, with about 6.5 million pigs in England and Wales, plus another million in Scotland and Northern Ireland, for the UK total to remain at around 7.5 million. Of this total some three-quarters of a million, or 10 per cent, will be sows and they should produce some 15 million slaughter pigs a year. This is a slightly better production rate than that of the early 1990s of just over 14 million because some improvement in pigs per sow a year is likely.

Table 17.1 Changes in number of holdings and total pigs by size groups and forecast for year 2000. England and Wales

	1970	1980	1990	2000 forecast
Holdings				
Number				
1 -99	44,834	13,791	7,615	5,700
100 -999	14,294	7,520	4,174	3,300
1,000 & over	725	1,662	1,818	2,000
Total	59,853	22,973	13,607	11,000
Percentage	%	%	%	%
1 -99	75	60	56	52
100 -999	24	33	31	30
1,000 & over	1	7	13	18
Total	100	100	100	100
Pigs				
Number	'000	'000	'000	'000
1 -99	1,230	339	160	110
100 -999	3,955	2,667	1,695	1,390
1,000 & over	1,223	3,602	4,536	5,000
Total	6,408	6,602	6,391	6,500
Percentage	%	%	%	%
1 -99	19.2	5.1	2.5	1.7
100 -999	61.7	40.4	26.5	21.3
1,000 & over	19.1	54.5	71.0	77.0
Total	100	100	100	100

Performance

Substantial improvements in performance were achieved between 1970 and 1990 but, because many factors have now reached a high standard, it is unlikely that the same rates of improvement can be maintained to 2000. Some forecasts compared with past performance are given in Table 17.2.

Table 17.2 Estimated average performance for the year 2000

	1970	1980	1990	2000 forecast
Breeding				
Litters per sow	1.95	2.14	2.26	2.30
Live pigs born per litter	10.3	10.3	10.6	10.8
Weaners per litter	8.6	8.8	9.1	9.3
Weaners per sow in herd[a]	16.8	18.9	20.6	21.4
Feed per weaner to 8 wks	95 kg	82 kg	79 kg	77 kg
Feed per weaner to 30 kg	123 kg	106 kg	103 kg	100 kg
Feeding from 8 wks				
Mortality percentage	3.8 %	2.5 %	2.2 %	2.0 %
Feed conversion rate	3.79	3.20	2.65	2.45
Feeding from 30 kg				
Mortality percentage	2.5 %	1.5 %	1.9 %	1.5 %
Feed conversion rate	4.13	3.38	2.82	2.60
Breeding and feeding				
Feed per kg deadweight[b]	5.5 kg	4.8 kg	4.1 kg	3.8 kg

(a) To eight weeks of age
(b) For 66 kg deadweight pigs

The number of litters per sow in herd a year increased steadily through the 1970s and 1980s as more herds changed to weaning earlier. Now most are weaning by three to four weeks of age and this opportunity of changing to produce more litters a year remains open to only a few breeders. Further improvement to average performance during the 1990s is therefore likely to be only marginal. Litter size at birth remained almost constant at 10.3 pigs during the 15 years to 1983, and then during the next five years improved to 10.6, where it stayed until the final year of the scheme. Not much change is expected in the next few years but further increases are possible later on.

A recent importation of the prolific Meishan breed from China, for cross breeding experiments by some of the breeding companies, should eventually lead to the availability of stock capable of producing larger litters than at present. The original breed has drawbacks, however, in the form of very fat and poorly conformed carcases. The obvious aim, therefore, is to combine the prolificacy of the Meishan with the carcase quality of European breeds. Reports of some of the initial cross breeding results are encouraging. The gradual development of prolific stock will take time and several years could elapse before there is any major increase in the national average. By the year 2000 only the small improvement to 10.8 live pigs born per litter is expected.

Mortality in the breeding stage is also not expected to change and is forecast to remain at about 1.5 deaths per litter. With slightly larger litters at birth and the same mortality rate, the number of weaners surviving to the eight week stage should, on average, increase to 9.3 per litter and maintain the steady increase achieved over many years. The small improvement in litters per sow a year and in weaners reared per litter should result in an average of 21.4 weaners per sow in herd in the year 2000. This increase will be slower than in previous decades which benefited from the move to earlier weaning.

Future feed requirements per weaner produced will be influenced by the proportion of outdoor herds. Sows kept outdoors use more feed than indoor sows and, if their numbers increase more than expected, the total forecast for the year 2000 of 77 kg per weaner (to eight weeks of age) could be optimistic. This quantity allows for some increase in the number of outdoor units and shows a smaller improvement during the 1990s than previously. Here the major factor is the number of weaners produced per sow, over which to divide the sow feed, and the increase each decade has been declining as the average quantity has reached levels whence further improvement becomes more difficult. A similar rate of improvement in feed per weaner is forecast for herds producing 30 kg weaners.

Feeding stock mortality from eight weeks of age is forecast to be 2.0 per cent in the year 2000, which maintains the steady rate of improvement since the early 1970s. The feed conversion rate is by far the most important factor of production, and improvement during the 1990s is expected to continue but at a slower rate than hitherto and could be down to 2.45 by 2000. When starting with 30 kg weaners, the mortality rate here is expected to return to a level of about 1.5 per cent and feed conversion rates improve to average around 2.6. The availability and use of cheap by-products and their price relationship with cereals could make a difference to conversion rates. By-products usually give poorer rates but having the advantage of being cheap, they often provide low-cost liveweight gain. The 2.45 conversion rate estimated for the year 2000 relates to pigs of 66 kg deadweight and assumes a similar spread of types of feed to that used in 1990. Should more by-products be used the rate could be higher, and conversely if usage falls.

For breeding and feeding combined the total quantity of feed used per kg deadweight of pigs produced should also continue to improve but again not at the rate of previous decades. The 1970s and 1980s showed substantial improvements of 0.7 kg of feed per kg deadweight, whereas the 1990s, from a much higher standard to start with, are forecast to improve from 4.1 to 3.8 kg. The estimated feed requirement of 3.8 kg for the year 2000 would be 30 per cent less than the amount used in 1970.

Variation

As in previous years the wide variation of individual herd results around the average will continue. Estimates of the likely variation in performance has been derived from the 1991 results by using quartile distribution separately for each factor. One in four herds is expected to perform better than the upper quartile on each measure and one in four worse than the lower quartile. Not all factors follow the same trend. For example, for weaners per sow in herd, performance improves as the size of the number increases, while with feed conversion rates smaller numbers are best. In other words, the upper quartile always gives the best results.

A quarter of breeding herds should achieve more than 23.0 weaners per sow in the year 2000 and a quarter should produce less than 20.3 weaners. For the central half of herds, in

Table 17.3 Estimated variation in performance for the year 2000 (each factor separately)

	Average	Lower[a] quartile	Upper[a] quartile
Breeding			
Litters per sow in herd	2.3	2.2	2.4
Live pigs born per litter	10.8	10.3	11.3
Weaners per litter	9.3	8.9	9.8
Weaners per sow in herd	21.4	20.3	23.0
Sow feed used per sow	1.3 t	1.4 t	1.2 t
Sow feed per weaner	61 kg	65 kg	55 kg
Piglet feed per weaner 8 wks	16 kg	17 kg	15 kg
Piglet feed per weaner 30 kg	39 kg	43 kg	36 kg
Feeding from 8 wks			
Daily liveweight gain	.68 kg	.61 kg	.76 kg
Mortality percentage	2.0 %	2.6 %	1.0 %
Feed conversion rate	2.60	2.65	2.22
Feeding-only from 30 kg			
Daily liveweight gain	.71 kg	.64 kg	.79 kg
Mortality percentage	1.5 %	2.0 %	.8 %
Feed conversion rate	2.58	2.80	2.34
Breeding and feeding combined			
Feed per kg deadweight[b]	3.8 kg	4.1 kg	3.5 kg

(a) Estimates based on quartiles from the 1991 results. One in four herds are expected to perform better than the upper quartile on each measure and one in four worse than the lower quartile.

(b) For 66 kg deadweight pigs.

terms of performance, feed requirements for sows and piglets are likely to extend from 70 kg to 82 kg per eight week old weaner and from 91 kg to 108 kg for producing 30 kg weaners. The equivalent ranges in conversion rates for feeding stock are projected to be from 2.22 to 2.65 for herds finishing home-bred weaners from eight weeks of age and from 2.34 to 2.80 for feeder-only herds starting with 30 kg weaners. For breeding and feeding combined, the quantity of feed used per kg deadweight of pigmeat for one in four herds should be below 3.5 kg and for one in four above 4.1 kg.

Feeding costs

In recent years it has been difficult enough to forecast feed prices with any confidence for just one year ahead. To do so for several years ahead, with all the uncertainties from likely changes in CAP and GATT legislation, is even more difficult. At this stage it seems best to use a range of prices per tonne to indicate feed costs per kg deadweight and per pig, hoping that one of them may be near the mark come the year 2000. Table 17.4 shows feed costs for forecast average performance with upper and lower quartiles, at prices varying from £120 to £180 per tonne.

Table 17.4 Estimated feed costs at varying prices per tonne (breeding and feeding)

	Upper quartile	Average	Lower quartile
Kg of feed used	3.5	3.8	4.1
Cost per kg deadweight	p	p	p
Feed @ £120 per tonne	42.0	45.6	49.2
Feed @ £140 per tonne	49.0	53.2	57.2
Feed @ £160 per tonne	56.0	60.8	65.6
Feed @ £180 per tonne	63.0	68.4	73.8
Cost per 66 kg dwt pig	£	£	£
Feed @ £120 per tonne	27.72	30.10	32.47
Feed @ £140 per tonne	32.34	35.11	37.88
Feed @ £160 per tonne	36.96	40.13	43.30
Feed @ £180 per tonne	41.58	45.14	48.71

The overall price of feed per tonne for breeding and feeding herds must include all feed for sows, piglets and feeders to slaughter weight. At an average price of £160 per tonne, feed costs with a upper quartile standard of performance would be 56p per kg deadweight, or £36.96 per 66 kg pig, while a lower quartile standard would cost 65.6p per kg, or £43.30 per pig. The difference between the two standards of performance amounts to 9.6p per kg deadweight, or £6.34 per pig.

Labour and other costs

In 1991, labour costs for breeding and feeding herds producing 66 kg deadweight pigs averaged £9.54 per pig, other costs £10.37 and stock depreciation £1.02; a total of £20.93. The total varied from £17 per pig for the upper quartile to £25 for the lower quartile. Inflation will cause these costs to rise but some savings are likely from continued improvement in productivity. By the year 2000, in current terms they could perhaps reach on average £25 per pig but vary between £20 and £30 per pig. In real terms of money at 1991 buying power, however, labour and other costs are likely to fall and could be lower than at present. This also applies to feed costs, so in real terms total costs of pig production should continue falling. The changing structure of costs could mean that labour and other costs will form a larger share of the total than they have done previously. Pig producers are unlikely to be any better off as a result of falling costs in real terms because prices paid for pigs are sure to follow the same trend.

Margins

Over the five year period between 1987 and 1991 herds in the Cambridge scheme achieved on average a margin of £9 per £100 output, before charging interest on capital invested, but this ranged from a deficit of nearly £2 in 1988 to a surplus of £20 in 1989. The five year period had two financially good years for pig producers and two poor years, with one just below average. It seems quite likely that fluctuations of this magnitude will contin-ue in future years and that an average of about £10 per £100 output could be the standard. As the number of pig producers declines, one might expect the variation in profitability between herds to reduce. So far, this has not happened and the range in recent years has remained as great as ever. In the five years

1987-91 the most profitable herds averaged £21 surplus margin per £100 output and the least profitable showed a deficit of over £13. This level of variation in profitability around the average by individual herds could still exist in the year 2000.

Outlook

The outlook for pig producers in the years ahead seems likely to include more controls and legislation on the way pigs are kept. Practically all producers treat their pigs kindly and keep them in reasonable comfort. Most stay abreast of modern developments and, when profits permit, invest in better housing and equipment. Production efficiency is continually improving and producers are well aware that failure to update their management and technical skills will soon leave them with an unprofitable unit. Further restrictions seem unnecessary to the majority of today's pig-keepers and only come about from the methods used by a tiny minority which upset certain members of the general public. These few cases are often highlighted by the media, usually to the detriment of all producers through adverse reaction of some consumers. Perhaps more should be done to publicise the good standards met by most pig producers.

In recent years consumption of pork has been increasing and is forecast to reach record levels in 1993 and 1994. Bacon consumption has been falling but should remain steady for the next few years. As pork is competitively priced with beef and lamb, the demand for pork should remain firm but care is needed not to over-produce as it will then be difficult to prevent a price slump until supplies are reduced. Exports of pork have increased considerably in recent years and are now expected to match or exceed imports in future years. There seems little prospect of increasing the demand for bacon, either home-produced or imported. Pig production costs in the middle to late 1990s should be contained, apart from the odd hiccup, as reduced support for cereals should lower feed prices and at least offset the rise in labour and other costs.

The difference between production costs and returns from pigs sold is, of course, the margin before charging for interest on capital. Overall margins during the 1990s are forecast to average £10 per £100 output but the most profitable herds are likely to make £20 or more, while the least profitable will lose money.

Individual producers will be rewarded according to how efficient they are. While it is difficult to improve an already efficient herd, the range of individual results shows that room for improvement exists for many.

Keeping records to measure performance achievement levels and attending to matters where there is scope for improvement are the keys to successful pig keeping. The best producers, achieving good results, have survived difficult periods in the past, and their unceasing efforts to improve still further place them in a excellent position to maximise income when pigs are profitable. In the long-term good producers should earn reasonable profits from pigs.

The termination of the Cambridge Pig Management Scheme in 1991 has ended an era of continuous monitoring by university departments of the fluctuating fortunes of pig producers and the standards of performance they achieved. Results of the scheme, published annually, provided the industry with reliable independent information to assist management and policy making decisions. Long-term changes were carefully documented to provide a sound historical record of achievement. The responsibility for the provision of any similar service now rests with others.

Appendix 1

Definition of Terms
as used by the Cambridge Pig Management Scheme

Averages
The output weighted average was used throughout unless stated.

Breeding stock
Included boars, sows and gilts for breeding, plus young pigs until eight weeks of age.

Feeding stock
Included all pigs over eight weeks of age which were not breeding stock. Gilts required for breeding were reared with feeding stock and transferred to breeding stock at approximately 90 kg liveweight.

Deadweight to liveweight conversion
Liveweight = (Deadweight^2 x -0.0028676) + (Deadweight x 1.55016) - 1.4066.

Stock valuations
Breeding stock purchased in the preceding six months were valued at cost and others at standard rates which were the approximate averages of cost and cull value. High priced stock were depreciated from cost to the standard value over two years to avoid excessive depreciation in the first full period. Home-bred breeding stock were valued by age up to the standard rate. Young pigs under eight weeks of age were valued by size or age and feeding stock by weight according to market prices at time of valuation.

Payments for boars
Subsidised boars were charged at full cost and the subsidies added to the value of weaners/stores purchased or sold.

Pigs purchased
Haulage and marketing charges were added to the cost of pigs purchased.

Pigs sold
Levies, haulage and marketing charges have been deducted from the value of pigs sold. Any condemned pigs for which no payment was received were classed as deaths, though weight and payment for partly condemned were included.

Feed
Purchased feed was charged at net cost and home-grown at the expected price for which similar quality could be sold in the month of use.

Labour
Included gross wages for pig managers, pigmen and assistants, plus any other farm labour, overtime pay, holidays, bonuses, employer's share of national insurance and any pension contributions, together with the value of any perquisites such as free housing. Any work carried out by the farmer was charged for at the Grade I rate of pay and family labour at the equivalent employed rate.

Other costs
Included charges for farm transport to cover the use of tractors, moving pigs on the farm and any other vehicles used in connection with pigs, veterinary costs, A.I., repairs and maintenance, power and water, purchased litter, baling home-grown straw, depreciation of equipment, depreciation (or rent) of buildings, pasture, insurance and other miscellaneous expenses.

Capital
The assessment of capital was intended to reflect the average amount required for the established herds in the scheme. The sum required to start a new herd at present day prices would be much higher. Breeding stock were included at the average of the opening and closing valuations. Buildings, including those rented, were allowed for together with equipment. New buildings were charged at cost. Other buildings were included at current (replacement) value in relation to present condition. The cost of any major alterations were taken into account. All buildings were revalued annually to allow for inflation. Working capital was also included to cover feed, labour and other costs over the expected time pigs remain on the farm.

Current and real terms
Costs and returns were first measured in current terms at the rates actually paid or received. When making comparisons between years it is usually more

meaningful to adjust these values, to the level of the latest year, to take account of changes in the purchasing power of the pound due to inflation. All values were reflated by the Retail Price Index (RPI) to the value of the pound for the year chosen for the base period, so that from one year to another costs can be compared in what is known as real terms.

Interest charge
No charge was included for interest on capital unless stated.

Appendix 2

Explanation of Calculations

1. General

(a) Livestock output
The sum of the value of pigs sold and the closing valuation, less the sum of the cost of pigs purchased and the opening valuation.

(b) Costs related to output
The total cost of feed, labour and other costs, together with the resulting surplus (or deficit) margin, was shown as a percentage of livestock output.

(c) Feed preparation costs
The cost of milling and mixing (and pelleting where applicable) were added to feed costs at the end of each period. Contract milling and mixing was charged at cost. When undertaken by the farmer the charge included depreciation (10 per cent) of the current value of buildings and equipment (including storage), Pharmaceutical Society registration fees, repairs, replacements, running costs, labour and an allowance for milling loss. These were charged according to each farm's own costs and not at a standard figure. If feed was also prepared for other livestock, only the pigs' share was charged.

2. Breeding (physical)

(a) Number of sows in herd
The mean of the monthly average number of sows and in-pig gilts (but not unserved gilts) over the recording period.

(b) **Litters per sow**
The total number of litters born during the period, divided by the number of sows in herd (2a).

(c) **Pigs born per litter**
The total number of pigs born (alive when first seen) divided by the number of litters.

(d) **Weaners per litter**[1]
The total number of pigs born, less deaths under eight weeks of age, divided by the number of litters.

(e) **Weaners per sow**[1]
The total number of pigs born, less deaths under eight weeks of age, divided by the number of sows in herd (2a).

(f) **Weight of weaners**[1]
The total liveweight of weaners leaving the breeding herd at eight weeks of age divided by the number of weaners.

(g) **Quantity of feed**[1]
The total quantity of sow feed, plus any other feeds (milk, potatoes, etc.) converted to meal equivalent (Appendix 4), divided by the following:
(i) Number of sows in herd (2a)
(ii) Total number of weaners produced (births less deaths)[1]

(h) **Quantity of piglet feed**
The total quantity of piglet feed divided by the total number of weaners produced.

(i) **Culled sows**
The total number of sows sold was shown as a percentage of the number of sows in herd (2a).

---

[1] These factors were also calculated to the time of sale for the group of breeding herds selling weaners.

3.   Breeding (financial)

(a)   Costs per weaner[1]
(i) **Feed**
The total cost of all breeding stock feed (including piglet feed) divided by the total number of weaners produced.
(ii) **Labour**
The total cost of labour allocated to breeding stock divided by the total number of weaners produced.
(iii) **Other**
The total other costs allocated to breeding stock divided by the total number of weaners produced.
(iv) **Stock depreciation**
The opening valuation of boars, sows and gilts, plus purchases and transfers in, less the closing valuation and sales, divided by the total number of weaners produced.

(b)   **Cost of sow meal per tonne**
The total net cost of all sow meal (compounds and own mixed both purchased and home-grown), including feed additives and preparation charges (1c), divided by the quantity.

(c)   **Cost of piglet meal per tonne**
The total cost of all piglet meal (compounds and own mix) including any feed additives and feed preparation charges (1c), divided by the quantity.

4.   Feeding stock (physical)

(a)   **Number of feeders in herd**
The mean of the monthly average number of feeding stock over the recording year (or half-year).

(b)   **Liveweight of pigs produced**
The total liveweight of all feeding stock sold and transferred to breeding stock, divided by their number.

(c)   **Liveweight of pigs brought in**
The total liveweight of all weaners and stores purchased and transferred from breeding stock, divided by their number.

(d)   **Liveweight gain**
The sum of the total liveweight of pigs produced (4b) and the closing valuation, less the sum of he total liveweight of pigs brought in (4c) and the opening valuation.

(e)   **Feed conversion rate**
The total quantity of feeding stock feed, including other feeds (milk, cereal starch, etc.) converted to meal equivalent (see Appendix 4), divided by the liveweight gain (4d).

(f)   **Mortality**
The number of feeding stock deaths as a percentage of the number of pigs brought in (4c), plus the number of feeding stock in the opening valuation, less the number in the closing valuation.

(g)   **Daily liveweight gain**
Total liveweight gain (4d), divided by the number of days in the recording period, then divided by the number of feeders in herd (4a).

5.   **Feeding stock (financial)**

(a)   **Costs per kg liveweight gain**
(i) Feed
The total cost of all feeding stock feed divided by the liveweight gain (4d).
(ii) Labour
The total cost of labour allocated to feeding stock, divided by the liveweight gain (4d).
(iii) Other
The total other costs allocated to feeding stock, divided by the liveweight gain (4d).
(iv) Mortality charge
The average value of weaners and stores purchased and transferred to feeding stock, multiplied by the number of feeding stock deaths and divided by the liveweight gain (4d).

(b)   **Cost of meal per tonne**
The total net cost of all feeding stock meal (compounds and own mixed both purchased and home-grown), including feed additives and preparation charges (1c), divided by the quantity. Excluded other feeds.

(c)   **Cost of feed per tonne**
The total cost of all feeding stock feed including other feeds (milk, cereal starch, etc.), divided by the total quantity including other feeds converted to meal equivalent (see Appendix 4).

6.   **Pig weights and prices**

(a)   **Weight per pig purchased**
The total liveweight of pigs purchased in each category (piglets, weaners and stores) divided by the number purchased.

(b)   **Price per pig purchased**
The total cost of pigs purchased in each category (piglets, weaners and stores) including any haulage and procurement charges, divided by the number purchased.

(c)   **Weight per pig sold**
The total weight of pigs sold in each category (liveweight for piglets, weaners and stores, deadweight for porkers, cutters, baconers and heavy pigs), divided by the number sold.

(d)   **Price of pigs sold**
The total net price of pigs sold in each category, after deduction of haulage and marketing charges, divided by the following:
(i)   the number sold
(ii) the total deadweight (kg)

Appendix 3

Deadweight to liveweight conversion

Liveweight = (Deadweight^2 x -0.0028676) + (Deadweight x 1.55016) - 1.4066

Weight Conversion Table

Dead-weight kg	Live-weight kg	Killing out %	Dead-weight kg	Live-weight kg	Killing out %	Dead-weight kg	Live-weight kg	Killing out %
40	56.01	71.42	60	81.28	73.82	80	104.25	76.74
41	57.33		61	82.48		81	105.34	
42	58.65		62	83.68		82	106.42	
43	59.95		63	84.87		83	107.50	
44	61.25		64	86.06		84	108.57	
45	62.54	71.95	65	87.24	74.51	85	109.64	77.53
46	63.83		66	88.41		86	110.70	
47	65.11		67	89.58		87	111.75	
48	66.39		68	90.74		88	112.80	
49	67.66		69	91.90		89	113.84	
50	68.93	72.54	70	93.05	75.23	90	114.88	78.34
51	70.19		71	94.20		91	115.91	
52	71.45		72	95.34		92	116.94	
53	72.70		73	96.47		93	117.96	
54	73.94		74	97.60		94	118.97	
55	75.18	73.16	75	98.72	75.97	95	119.98	79.17
56	76.41		76	99.84		96	120.99	
57	77.64		77	100.95		97	122.00	
58	78.86		78	102.06		98	123.00	
59	80.07		79	103.16		99	124.00	
						100	125.00	80.00

Appendix 4

Meal equivalent conversion factors used
(quantity to equal 1 kg M.E.)

Skim milk	8.6	litre
Whole milk	4.45	litre
Whey	15.0	litre
Whey concentrate	2.5	litre
C starch	4.5	kg
Abracarb	3.3	kg
Abrapro	2.0	kg
Potatoes (raw)	5.0	kg
Potatoes (cooked)	4.0	kg
Liquid potato feed	8.5	kg
Yeast (brewers)	4.3	kg
Yeast (brewers) pressed	3.0	kg
Beer waste	7.2	litre
Sugar/fodder beet	5.0	kg
Vegetable waste	8.0	kg
Carrots	5.0	kg
Mangolds	10.0	kg

Appendix 5

Specimen forms

Monthly recording form

Stock valuation form

Labour and other costs form

Specimen of herd results

Amalgamated financial results 1991

Purchased and home-grown feedingstuffs used 1991

Monthly recording form

UNIVERSITY OF CAMBRIDGE

Agricultural Economics Unit
Department of Land Economy
Silver Street
Cambridge CB3 9EP

PIG MANAGEMENT SCHEME

Code Number _____ 726

Details for the month of *September* 19 *91*   Name ____ *SPECIMEN*

STOCK

Number at 1st of month	Number		Number at end of month	Number	
Boars	11	brought forward from previous return.	Boars	10	The total number of sows and gilts here should equal the number at 1st of month plus purchases and transfers in, minus sales and deaths.
Sows and in-pig gilts . . .	204		Sows and in-pig gilts . .	202	
Unserved gilts	20		Unserved gilts	16	
Pigs under 8 weeks . . .	685		Pigs under 8 weeks . . .	674	
Feeders over 8 weeks . . .	1192		Feeders over 8 weeks . . .	1204	

Purchases	Number	Total Price £		Sales	Number	Total Price £	Total Livewt kg or	Total Deadwt kg
Boars	1	600		Boars	2	210		
Sows				Sows	15	1650		
Gilts	6	990	Total Livewt kg	Gilts				
Weaners				Weaners				X
Stores				Stores				X
				Porkers	30	1425	2070	
				Cutters	312	18944		20592
Farrowings				Baconers				
Litters	40	X		Heavy pigs				
Pigs born alive	421			Feeder culls	4	153		180
				Deaths:- Boars				
				Sows and gilts	1			
				Pigs under 8 weeks . . .	62			
				Feeders over 8 weeks . . .	8			

TOTAL 2540 ←——— (These should agree) ———→ TOTAL 2540

Average age at weaning	Days 26	Average liveweight of weaners at 8 weeks	kg 18	Own bred stock transferred from feeders to breeders		Number	kg livewt
					Boars		
					Gilts	4	360

FEED

Own mixed

Composition of mixtures	Price/Value per tonne £	Quantity per mix or per tonne				
		Sow kg	Rearer kg	Grower kg	Finisher kg	kg
Home-grown ingredients						
Wheat	106	450	450	400	385	
Barley	100	300	150	275	300	
Purchased ingredients 1						
Wheat						
Barley						
Middlings	102	100	100	75	50	
Soya bean	150		175	175	200	
Fish meal	355		50	35	25	
Sow concentrate	260	150				
Milk powder	510		25			
Fat premix	198		25	20	20	
Minerals/vits	800		25	17		
Minerals/vits	300				18	
Additives	1100			3	2	
Quantity issued 2. Breeding stock.	mixes tonnes	16				
Piglets	mixes tonnes		2			
Feeders	mixes tonnes		1	12	40	

Purchased Compounds

Type	Net Price £	Quantity		
		Breeders tonne	Piglets tonne	Feeders tonne
Sow nuts	@ 148	8		
Starter	@ 450		½	
Creep	@ 300		1½	
Rearer	@ 210		2	1
Grower	@ 170			6

Other feeds *

Type	Net Price £	Quantity		
		Breeders	Piglets	Feeders
C Starch	@ 13			10 t

1. Include minerals, feed additives, etc.
2. State mixes or tonnes (delete as appropriate)

* Include potatoes, milk, C. starch, etc. and state whether home-grown (HG) or purchased (P).

Stock valuation form

Agricultural Economics Unit	PIG MANAGEMENT SCHEME	Code number *726*
Department of Land Economy	STOCK VALUATION	
University of Cambridge	Date *30 Sept 1991*	Name .... *SPECIMEN*

BREEDING STOCK	Number	Average value £	Total value £
Boars	*10*	*350*	*3,500*
Sows	*170*	*120*	*20,400*
In-pig gilts (purchased)	*20*	*165*	*3,300*
(own bred)	*12*	*100*	*1,200*
Unserved gilts (purchased)	*10*	*165*	*1,650*
(own bred)	*6*	*80*	*480*
Total sows and gilts	*218*		*27,030*
Suckling pigs	*312*	*18*	*5,616*
Weaners under 8 wks	*362*	*22*	*7,964*
Total pigs -8 wks	*674*		*13,580*

FEEDING STOCK	Number	Average liveweight kg or lb *	Total liveweight kg or lb *	Average value £	Total value £
include all feeders over 8 wks of age	*203*	*24 kg*	*4,872 kg*	*28*	*5,684*
	163	*32*	*5,216*	*32*	*5,216*
	184	*40*	*7,360*	*36*	*6,624*
* delete as appropriate	*170*	*50*	*8,500*	*41*	*6,970*
	177	*60*	*10,620*	*46*	*8,142*
	165	*72*	*11,880*	*52*	*8,580*
if insufficient space, use reverse side of this form	*142*	*80*	*11,360*	*56*	*7,952*
Total feeders	*1,204*		*59,808*		*49,168*

FEED already recorded on monthly returns but not consumed by date of valuation

Description	Price per tonne £	Quantity Breeders	Piglets	Feeders
Sow nuts	*150*	*2 t*		
Rearer	*210*		*½ t*	*½ t*
Grower	*170*			*2 t*

Labour and other costs form

CAMBRIDGE PIG MANAGEMENT SCHEME

Labour & other costs
for six months ending _____

Code number _____
Name _____

	Costs £	Share of costs Breeders & piglets to 8 wks %	Feeders 8 wks & over %
1 LABOUR			
Farmer's own labour (_____ hours per week)			
Family labour (_____ " " ")			
Farm manager			
Pig unit manager			
Pigmen/women			
" "			
" "			
Others			
Secretarial			
Holiday relief			
Holiday pay			
Staff bonuses			
Housing			
Other perquisites			
National insurance (employer's share)			
" " (self-employed)			
Contract/casual work			
" slurry removal (share to labour)			
If labour includes any of the following tasks, please estimate the number of hours worked this six months.			
Milling & mixing ... ... _____ hours			
Maintenance – mill & mixer plant ... ... _____ hours			
" – buildings & equipment _____ hours			
New buildings & alterations ... ... _____ hours			
Haulage of pigs purchased/sold ... ... _____ hours			
Baling & carting straw ... ... _____ hours			
Cooking potatoes, swill, etc. ... ... _____ hours			

	£	%	%
2 FARM TRANSPORT & TRACTORS			
Used exclusively for pigs — Tractors £ — Other vehicles £			
Current value	x	x	x
Depreciation @ 12½% × ½...	x	x	x
Repairs & maintenance	x	x	x
Fuel & oil	x	x	x
Tax & insurance	x	x	x
_____ + _____ =			
If above includes any of the following tasks, please estimate percentage of time worked this six months.			
Milling & mixing ... ... _____ % ...			
Haulage of pigs purchased/sold ... ... _____ % ...			
Baling & carting straw ... ... _____ % ...			
Used occasionally for pigs			
Farm tractors _____ hours @ £ _____ per hour ...			
Farm vehicles _____ " @ £ _____ " " ...			
Car/pick-up journeys exclusively for pigs			

	Costs £	Breeders %	Feeders %

3 VET & MEDICINES (exclude VAT)
Veterinary service ... ...
        "   supplies* ... ...
        "   equipment ... ...
Disinfectants			

*If feed additives included, please state _____ kg	£ ____		
4 ARTIFICIAL INSEMINATION A.I. fees ... ... " equipment	£	%	% x x x

5 POWER & WATER
Electricity – total ... ...
      "   – used for purposes other than pigs ... ...
      "   – used for milling & mixing ... ...
      "   – remainder to the pig unit ... ...
Gas   ... ...
Water – mains ... ...
      "   – own well ... ...
	£
	_
	_
		%
		x
		x
		x
			%
			x
			x
			x

6 MISCELLANEOUS EXPENSES (exclude VAT) Small tools, brooms, shovels, forks, etc. ... ... Lamps & bulbs ... ... Markers, ear tags ... ... Protective clothing ... ... Laundry ... ... Pregnancy & genetic testing fees ... ... Subscriptions ... ... Recording fees ... ... Weighbridge fees ... ... Computer fees ... ... Consultancy fees ... ... Accountant's fees ... ... Insurance premiums ... ... Telephone, stationery, postage ... ... Pest control ... ... _____ ... ... _____ ... ... _____ ... ... _____	£	%	%

7 LITTER (used this six months)
Wood shavings ... ...
Purchased straw ... ...
Baling & carting straw _____ tonnes @ _____ ...
      "   "   "   _____ bales (ord) @ _____ ...
      "   "   "   _____ bales (big) @ _____ ...
      "   "   "   by contractor ... ...
 	
*If manure is of value, deduct sum from above costs
	£
	
	_
		%	%

Current value of baler and bale handling equipment	£ ____ pigs share ____ %		

	Costs £	Breeders %	Feeders %
8 MAINTENANCE — (exclude new buildings & alterations) (exclude tractors, mill & mixer)			
Buildings			
Equipment		.	
Electrical			
Plumbing			
Painting, creosoting, etc, ...			
... ...			

9 EQUIPMENT (exclude VAT)   £

	Costs £	Breeders £	Feeders £
Valuation of equipment (b.fwd) _____	x		
Purchase (please specify)			
_____ _____	x		
_____ _____	x		
_____ _____	x		
_____ _____	x		
_____ _____	x		
Less sales ... ... _____	x		
_____ @ 20% × ½ =			

Hire of equipment (please specify) _____

Contract slurry removal (share to equipment) ... ...

10 BUILDINGS   £

	£	£	£
Valuation of buildings (b.fwd) _____	x		
New buildings & alterations			
_____ _____	x		
_____ _____	x		
_____ _____	x		
_____ _____	x		
_____ _____	x		
Less buildings no longer used ... ... _____	x.		
_____ @ 10% × ½ =			

Rented buildings (half yearly charge) ... ...

11 CHARGE FOR PASTURE

	£	%	%
Rent of land _____ ha/acre @ £ _____ per ha/acre a yr			
Cultivations, seed, etc.			

12 HAULAGE (where not already allowed for on monthly returns — exclude VAT)

	Purchases £	Sales £		Sales £
Boars	_____ ...	_____	Porkers	_____
Sows	_____ ...	_____	Cutters	_____
Gilts	_____ ...	_____	Baconers	_____
Weaners	_____ ...	_____	Heavies	_____
Stores	_____ ...	_____	Feeder culls	_____

Feed collected by own vehicle _____ tonnes (type _____) @ £ _____ per tonne

13 GROUP MEMBERSHIP MARKETING LEVIES (where not already allowed for)

Payments made in this half year   £_____ Covers year/six months*

     "   received   "   "   "   £_____

*delete as appropriate

14 FEED PREPARATION (own milling & mixing)   £

Valuation of buildings, equipment & storage (b. fwd) ... ... _____

Purchases (please specify)

               _____ ... ... _____

               _____ ... ... _____

               _____ ... ... _____

Less sales of equipment ... ... _____

    Total ... ... _____

Costs of milling & mixing   £

Depreciation (10% of above total × ½) ... ... _____

Repairs & replacement parts ... ... _____

Labour* _____ minutes per tonne @ £_____ per hour ... ... _____

Power — mill & mix* _____ units per tonne @ _____p per unit ... ... _____

      " — pelleting* _____ " " " @ _____p " " ... ... _____

Milling loss*_____ kg per tonne ... ... _____

*If no change from last time, leave blank and tick here ☐

    Quantity of feed prepared for livestock other than this pig unit ... ... _____ tonnes

FEED PREPARATION (by contractor)   £

Cost per tonne — Breeders _____ tonnes @ £_____ per tonne ... ... _____

    " " " — Feeders _____ tonnes @ £_____ " " ... ... _____

Milling loss* _____ kg per tonne ... ... _____

Farm labour* _____ minutes per tonne @ £_____ per hour ... ... _____

*Where applicable

COOKING (potatoes, swill, etc.)   £

Value of equipment/plant ... £_____ Depreciation (10% x ½) ... ... _____

Labour _____ minutes per tonne @ £_____ per hour... ... ... _____

Fuel ... ... _____

15 CREDITS received in this six months (though part may apply to a previous period)

Contract bonus on pig sales ... ... £_____

Additional payments received for pigs sold ... ... £_____

Discounts received on breeding stock purchases — Boars £_____   Gilts £_____

Feed discounts/refunds (type of feed _____) ... ... £_____

Insurance payments received (reason _____) ... ... £_____

Other (please specify) _____) ... ... £_____

16 BOAR SUBSIDIES (where not already allowed for)

Boars purchased — amount received £_____ amount paid £_____

Boars sold   —   "   " £_____   "   " £_____

17 FEED RECORD ADJUSTMENT (where necessary)

Where feed recorded on monthly returns has been fed to more than one class of stock (e.g. the same feed to pigs under eight weeks of age and over eight weeks) please give details.

Weaner feed (type _____) used for pigs from _____ wks/kg to _____ wks/kg

Sow feed used for rearing gilts in the feeding herd _____ kg per pig

Comments

_____

_____

_____

_____

_____

Specimen of herd results

Cambridge Pig Management Scheme

No.	Specification	£	£	No	Specification	£	£
	OPENING VALUATION				CLOSING VALUATION		
9	Boars	3,060		10	Boars	3,500	
212	Sows & gilts	26,182		218	Sows & gilts	27,030	
628	Pigs under 8 weeks	12,524		674	Pigs under 8 weeks	13,580	
1,165	Feeders	48,522		1,204	Feeders	49,168	
2,014			90,288	2,106			93,278
	PURCHASES				SALES		
4	Boars	2,240		3	Boars	306	
56	Sows & gilts	8,680		72	Sows & gilts	7,560	
	Piglets				Piglets		
	Weaners				Weaners		
	Stores	___			Stores		
			10,920	286	Porkers	15,310	
				4,002	Cutters	258,849	
5,123	Pigs born alive				Baconers		
					Heavy pigs		
				21	Feeder culls	861	
			101,208				282,886
					Bonuses		1,200
	LIVESTOCK OUTPUT				DEATHS		
	(carried down)		276,156	611	Before 8 wks of age	x	
				88	After 8 wks of age	x	
				8	Boars & sows	x	
7,197	Totals		377,364	7197	Totals		377,364

	COSTS	£			Livestock output		£
	Meal (1,073 tonnes)	170,416			(brought down)		276,156
	Other feeds	1,640					
			172,056				
	Labour		39,347				
	Farm transport	3,187					
	Vet & vet supplies	4,236					
	A.I. fees	520					
	Power & water	7,262					
	Miscellaneous exps	4,845					
	Litter	2,830					
	Maintenance	4,509					
	Equipment charge	2,308					
	Buildings charge	12,135					
	Pasture charge	___					
	Total costs		41,832				
	Surplus (profit)		22,921		Deficit (loss)		
			276,156				276,156

Costs per £100 livestock output	£	Composition of £100 livestock output	£
Feed	62.30	Sales	102.87
Labour	14.25	Purchases	3.95
Other costs	15.15	Sales less purchases	98.92
Total ocsts	91.70	Valuation - changing values	-0.22
Margin	8.30	Valuation - changing number	+ 1.30
	100		100

QUANTITIVE DETAILS

BREEDING STOCK

Number of sows in herd*	204
Number of litters	407
Litters per sow in herd	2.30
Live pigs born per litter	10.9
Weaners per litter	9.6
Weaners per sow in herd	22.1
Liveweight of weaners**	18.2 kg
Culled sows percentage	35 %
Feed per sow	1.27
Sow feed per weaner**	57.4 kg
Piglet feed per weaner**	17.2 kg
Cost of sow meal per tonne	£144.60
Cost of piglet meal per tonne	£268.00
Compounds as % of total feed	49 %

Costs per weaner**	£
Feed	12.91
Labour	5.39
Other costs	5.53
Stock depreciation	1.06
Total breeding costs	24.89

FEEDING STOCK

Liveweight of pigs produced	87.3 kg
Liveweight of pigs brought in	18.2 kg
Daily liveweight gain	.66 kg
Mortality percentage	2.0 %
Feed conversion rate***	2.52
Cost of meal per tonne	£152.26
Cost of feed per tonne***	£150.40
Compounds as percent of total feed	28 %

Costs per kg liveweight gain	p
Feed	37.9
Labour	5.0
Other costs	5.6
Mortality charge	.7
Total feeding costs	49.2

Feed preparation costs per tonne	£
Depreciation	1.64
Repairs and fees	.58
Labour	2.52
Power	1.56
Milling loss	.85
Total costs	7.15

AVERAGE PIG PRICES AND WEIGHTS

PURCHASES

Piglets	liveweight		kg
	price per pig	£	
Weaners	liveweight		kg
	price per pig	£	
Stores	liveweight		kg
	price per pig	£	

SALES

Piglets	liveweight		kg
	price per pig	£	
Weaners	liveweight		kg
	price per pig	£	
Stores	liveweight		kg
	price per pig	£	
Porkers	deadweight		53 kg
	price per kg dwt		101 p
	price per pig		£53.53
Cutters	deadweight		66 kg
	price per kg dwt		98 p
	price per pig		£64.68
Baconers	deadweight		kg
	price per kg dwt		p
	price per pig	£	
Heavies	deadweight		kg
	price per kg dwt		p
	price per pig	£	

Per cent sold as Piglets	
Per cent sold as Weaners	
Per cent sold as Stores	
Per cent sold as Porkers	6.6 %
Per cent sold as Cutters	92.9 %
Per cent sold as Baconers	
Per cent sold as Heavies	.5 %
Per cent sold as Culls	

* Monthly average (includes in-pig gilts)
** At 8 weeks of age
*** Includes any other feed converted to meal equivalent

No charge has been included for interest on capital.
Home-grown feeds have been charged at estimated market value in month of use.

Amalgamated financial details 1991

No.	Specification	£	£	No.	Specification	£	£
	Opening Valuation:				**Closing Valuation:**		
1,063	Boars	342,635		1,036	Boars	345,010	
21,222	Sows and gilts	2,622,593		20,691	Sows and gilts	2,524,006	
65,507	Pigs under 8 weeks	1,291,477		65,445	Pigs under 8 weeks	1,302,813	
111,542	Feeding stock	4,797,558		106,546	Feeding Stock	4,326,518	
199,334			9,054,263	193,718			8,498,347
	Purchases:				**Sales:**		
427	Boars	227,166		457	Boars	49,227	
5,989	Sows and gilts	896,756		16,483	Sows and gilts	1,695,359	
			1,123,922				1,744,586
15,357	Piglets	373,693		9,646	Piglets	255,445	
38,332	Weaners	1,242,268		63,386	Weaners	1,967,137	
62,198	Stores	2,178,831		83,633	Stores	2,900,304	
			3,794,792	76,148	Porkers	4,230,433	
				165,424	Cutters	10,631,917	
467,509	Pigs born alive	×		99,186	Baconers	6,764,138	
				331	Heavy pigs	22,831	
	Bad debts**		8,600	2,578	Casualties and culls	131,612	
			13,981,577				26,903,817
					Bonuses		61,888
	Livestock Output						
	Carried down		23,227,061				
					Deaths:		
				67,790	Before 8 weeks of age		×
				9,514	After 8 weeks of age		×
				850	Boars and sows		×
789,146			37,208,638	789,146			37,208,638

No.	Specification	£	£	No.	Specification	£	£
	Costs:						
	Meal (93,576 tonnes)	15,112,824			Livestock Output		
	Other feeds	239,967			Brought down		23,227,061
			15,352,791				
	Labour – hired	2,690,367					
	family	933,361					
			3,623,728				
	Farm transport	329,070					
	Vet and vet. supplies	405,664					
	A.I. fees	42,623					
	Power and water	613,901					
	Miscellaneous expenses	462,112					
	Litter	263,055					
	Maintenance	407,580					
	Equipment and fittings	223,351					
	Charge for buildings	1,100,860					
	Charge for pasture	33,804					
			3,882,020				
	Total costs	22,858,539					
	Surplus (Profit)*	368,522			Deficit (Loss)*		–
		23,227,061					23,227,061

* No charge has been made for interest on capital. Home-grown feeds have been valued at estimated market value in month of use.
** The amount owed to one producer for finished pigs bought by a company that ceased trading.

Purchased and home-grown feedingstuffs used 1991

Purchased concentrates	Total quantity tonne	%	Total value £	Av price per tonne £
Compounds - sows	16,410	16.9	2,460,692	149.95
" - feeders (over 8 wks)	24,795	25.7	4,158,815	167.74
- piglets (to 8 wks)	5,825	6.0	1,650,309	283.30
Concentrate mixes	2,856	2.9	728,872	255.20
Oil and fats (incl. full fat proteins)	1,242	1.3	227,603	183.25
Wheat	2,532	2.6	316,476	124.97
Barley	2,030	2.1	244,508	120.43
Cereal substitutes	445	.5	44,473	100.01
Milling offals	4,448	4.6	469,694	105.61
Soya bean	6,052	6.3	878,533	145.15
Fish meal	1,068	1.1	317,663	297.57
Milk powder	34	-	18,465	539.19
Maize	17	-	4,035	234.89
Micronised and cooked cereals	329	.3	52,253	158.59
Beans and peas	390	.4	54,268	139.09
Molasses	180	.2	17,996	99.89
Other proteins	442	.5	48,673	110.18
Additives (incl. minerals & vitamins)	748	.8	386,248	-
Home-grown concentrates				
Wheat	15,147	15.6	1,756,747	115.98
Barley	8,065	8.3	867,717	107.59
Oats	17	-	1,846	108.30
Rye	338	.3	39,198	116.01
Beans and peas	166	.2	19,090	114.66
Feed preparation costs			348,649	7.19
Total meal	93,576	96.6	15,112,824	161.50
Other feed*				
Milk and milk products	670	.7	47,973	71.57
Wheat starch	2,292	2.4	174,379	76.12
Other by products	164	.2	12,736	77.62
Liquid potato feed	8	-	969	116.76
Beer waste	17	-	518	30.03
Fodder beet	81	.1	3,392	41.86
Total all feed	96,808	100	15,352,791	158.59

*Quantity and price per tonne 'converted' to meal equivalent

267

Index

Farming Press

BOOKS

Outdoor Pig Production   Keith Thornton
The details of how to plan, set up and run a unit.

Pig Diseases   David Taylor
A detailed technical reference book about pig diseases for the veterinary surgeon and pig unit manager.

Growing and Finishing Pig   P. R. English, V. R. Fowler, S. H. Baxter & W. J. Smith
Explores in detail the many interlinked factors controlling the efficiency of pig growth from weaning to slaughter.

Pig Ailments   Eddie Straiton
The visual signs of ailments with clear details of treatment and prevention.

Pigman's Handbook   Gerry Brent
Describes in detail the best routines for the day-to-day running of a pig farm.

Farming Press, which is part of the Morgan-Grampian Farming Press Group, also publishes *Pig Farming*, Britain's leading magazine for pig farmers. For a specimen copy of this, please contact the address on the facing page.

Books and Videos

VHS VIDEOS

Practical Outdoor Pig Production **Keith Thornton**
An overview showing the main management considerations.

Lamb Survival **David C. Henderson**
Illustrates essential techniques for both lowland and hill shepherds seeking to reduce lamb losses.

Footcare in Cattle:
Hoof Structure and Trimming **Roger Blowey**
Roger Blowey first analyses hoof structure and horn growth using laboratory specimens, then demonstrates trimming.

Poultry at Home **Victoria Roberts**
An introduction to poultry management and health featuring 50 pure breeds.

Harnessed to the Plough **Roger & Cheryl Clark with Paul Heiney**
Roger and Cheryl Clark demonstrate a year of contemporary horsedrawn cultivations and harvesting on their Suffolk farm. Additional commentary by Paul Heiney.

For more information or for a free illustrated catalogue of all our publications please contact:

Farming Press Books & Videos, Wharfedale Road
Ipswich IP1 4LG, United Kingdom
Telephone (0473) 241122   Fax (0473) 240501

D